ECONOMICS OF AGRICULTURAL CROP INSURANCE: THEORY AND EVIDENCE

NATURAL RESOURCE MANAGEMENT AND POLICY

Editors:

EDITORIAL STATEMENT

There is a growing awareness to the role that natural resources such as water, land, forests and environmental amenities play in our lives. There are many competing uses for natural resources, and society is challenged to manage them for improving social well being . Furthermore, there may be dire consequences to natural resources mismanagement. Renewable resources such as water, land and the environment are linked, and decisions made with regard to one may affect the others. Policy and management of natural resources now require interdisciplinary approach including natural and social sciences to correctly address our society preferences.

This series provides a collection of works containing most recent findings on economics, management and policy of renewable biological resources such as water, land, crop protection, sustainable agriculture, technology, and environmental health. It incorporates modern thinking and techniques of economics and management. Books in this series will incorporate knowledge and models of natural phenomena with economics and managerial decision frameworks to assess alternative options for managing natural resources and environment.

Thus far, the series addressed natural resource issues explicitly. This book, by focusing on crop insurance, provides an insight to the management of an industry, agriculture, that relies heavily on natural resources. The book emphasizes (a) the importance of uncertainty in natural resource systems and the design of policy tools to address it, (b) the important role inventories play in management of resources, (c) the role of incentives to correct market failure, (d) the limited capacity of government interventions, and (e) the difficulties that arise in policy formulations because of imperfect and asymmetric information.

The Series Editors

Previously Published Books in the Series

Russell, Clifford S. and Shogren, Jason F.:
Theory, Modeling and Experience in the Management of Nonpoint-Source Pollution
Wilhite, Donald A.:
Drought Assessment, Management, and Planning: Theory and Case Studies

ECONOMICS OF AGRICULTURAL CROP INSURANCE: THEORY AND EVIDENCE

Edited by

Darrell L. Hueth
University of Maryland

William H. Furtan
University of Saskatchewan

Kluwer Academic Publishers
Boston/Dordrecht/London

Distributors for North America:
Kluwer Academic Publishers
101 Philip Drive
Assinippi Park
Norwell, Massachusetts 02061 USA

Distributors for all other countries:
Kluwer Academic Publishers Group
Distribution Centre
Post Office Box 322
3300 AH Dordrecht, THE NETHERLANDS

Library of Congress Cataloging-in-Publication Data
Economics of agricultural crop insurance : theory and evidence /
edited by Darrell L. Hueth, William H. Furtan.
p. cm. -- (Natural resource management and policy)
Includes index.
ISBN 0-7923-9435-6 (acid free paper)
1. Insurance, Agricultural--Crops--Developing countries.
I. Hueth, Darrell L. II. Furtan, W. H. (W. Harley) III. Series.
HG9968.E27 1994
368.1'2--dc20 93-47214
CIP

Printed on acid-free paper.

Printed in the United States of America

Contents

III Applications and Policy Studies

Contributing Authors

SERGIO ARDILA is an economist with the Environmental Protection Division of the Inter-American Development Bank in Washington, D.C.

LINDA CALVIN is an agricultural economist with the Economic Research Service, U.S. Department of Agriculture in Washington, D.C.

HARTLEY FURTAN is professor of agricultural and resource economics at the University of Saskatchewan at Saskatoon, Saskatchewan, Canada.

BRUCE GARDNER is professor of agricultural and resource economics at the University of Maryland College Park. During the period that this research was conducted he was on leave from the University of Maryland as Assistant Secretary for Economics, U.S. Department of Agriculture in Washington, D.C.

JOSEPH GLAUBER is Deputy Assistant Secretary for Economics in the Consumer Economics Division of the Economic Research Service, U.S. Department of Agriculture in Washington, D.C.

KEITH HAYWARD is an economist with the Saskatchewan Crop Insurance Corporation.

JULIE A. HEWITT is assistant professor of agricultural and resource economics at Montana State University in Bozeman, Montana.

JOHN HOROWITZ is assistant professor of agricultural and resource economics at the University of Maryland College Park.

DARRELL HUETH is professor and chair of agricultural and resource economics at the University of Maryland College Park.

ROBERT INNES is associate professor of agricultural and resource economics at the University of Arizona in Tucson, Arizona.

RICHARD E. JUST is professor of agricultural and resource economics at the University of Maryland College Park.

GIANNIS KARAGIANNIS is a research assistant at the University of Saskatchewan at Saskatoon, Saskatchewan, Canada.

HOWARD LEATHERS is associate professor of agricultural and resource economics at the University of Maryland College Park.

ERIK LICHTENBERG is associate professor of agricultural and resource economics at the University of Maryland College Park.

MARIO MIRANDA is assistant professor of agricultural and resource economics and rural sociology at The Ohio State University in Columbus, Ohio.

JOHN QUIGGIN was visiting associate professor of agricultural and resource economics at the University of Maryland College Park at the time of this work. He is currently senior research fellow of economics at the Research School of Social Sciences, Australian National University, Canberra, Australia.

ANDREW SCHMITZ is professor and chair of agricultural and resource economics at the University of California, Berkeley, California.

RONALD A. SCHONEY is professor of agricultural economics at the University of Saskatchewan at Saskatoon, Saskatchewan, Canada.

DALE SIGURDSON is Senior Analyst for the Saskatchewan Department of Finance.

RICKIE SIN is an economist with the Saskatchewan Department of Finance.

JULIE STANTON is a graduate research assistant of agricultural and resource economics at the University of Maryland College Park.

JULIA S. TAYLOR is a professional research associate with the University of Saskatchewan at Saskatoon, Saskatchewan, Canada.

WARD WEISENSEL is research associate of agricultural and resource economics at the University of Saskatchewan at Saskatoon, Saskatchewan, Canada.

BRIAN D. WRIGHT is associate professor of agricultural and resource economics at the University of California, Berkeley, California.

Acknowledgements

The publication of this book completes a project that began with a chance meeting of Grant Devine, Premiere of the Province of Saskatchewan, and William Donald Schaefer, Governor of the State of Maryland where they discussed common problems and expressed a desire to work together towards solutions. The Honorable Wayne A. Cawley, then Secretary of Agriculture, of the State of Maryland later asked the University of Maryland College Park about possible areas of mutual research interest. Conversations between the Editors of this book identified multiple peril crop insurance as an area of common discontent among growers and program administrators. Dr. Andrew Schmitz, who was at the time just completing a visiting professorship in the Department of Agricultural Economics at the University of Saskatchewan, expressed interest in Berkeley also being involved in the program. Thus, the pythagorean research program between College Park, Maryland, Saskatoon, Saskatchewan, and Berkeley, California was initiated and on April 2 and 3 of 1990, early drafts of research papers were presented at a Conference for the Improvement of Agricultural Crop Insurance held in Regina, Saskatchewan.

We thank Rulon Pope of the Department of Economics, Brigham Young University, Jock Anderson of the World Bank and Pinhas Zusman of Hebrew University of Jerusalem for providing valuable remarks and insightful comments on the early drafts of these papers at the conference. Thanks also to Nancy Quiggin for early editorial assistance with this project and Lien Trieu and Marie Klotz for top flight word processing services. Financial support from the Maryland Agricultural Experiment Station, Maryland Cooperative Extension Service, the Maryland Department of Agriculture and the Province of Saskatchewan is duly acknowledged.

This project would not have been possible had it not been for the high level of cooperation and support of the Commodity Economics Division of the Economic Research Service of the United States Department of Agriculture and its Director, Patrick O'Brien. In Saskatchewan, Deputy Minister of Agriculture,

Mr. Stewart Kramer provided both financial and moral support to the project. In Maryland, Mr. Robert L. Walker, then Deputy Secretary of Agriculture, provided essential coordination and liaison for all parties involved in the project.

ECONOMICS OF AGRICULTURAL CROP INSURANCE: THEORY AND EVIDENCE

Chapter 1

Introduction and Overview

D.L. HUETH[1] and H. FURTAN[2]

Agricultural producers in most countries make their land allocation and production decisions in an environment characterized by uncertainty with respect to both prices and yields. The attendant instability of farm income has generated pressure in many countries to devise programs and policies which will have a stabilizing effect on income. Government subsidized crop insurance has been used by a number of developed countries as a mechanism to reduce farm income instability by reducing yield risks. This book provides an in-depth analyses and evaluation of government provided crop insurance in two developed countries, the United States and Canada, but also summarizes experiences from other countries including Australia, Sweden and Japan.

Rather than looking at the single issue of the overall performance of government provided crop insurance, the approach taken here is to evaluate crop insurance programs from a number of perspectives for the U.S. and Canada. What has the effect of these programs been on environmental quality? How have the programs been influenced by agricultural credit markets? Do the commonly advanced theoretical reasons as to why the private insurance markets have not developed, mainly moral hazard and adverse selection, in fact, really explain the observed events? Can mechanisms be designed that mitigate moral hazard and adverse selection which will be more efficient and obtain broader participation at lower cost? Are there other income stabilizing programs which are more cost effective and politically feasible?

---

[1]Department of Agricultural and Resource Economics, University of Maryland, College Park, MD 20742, USA.

[2]Department of Agricultural and Resource Economics, University of Saskatchewan, Saskatoon, Saskatchewan, Canada.

Significant new information on each of these questions can be found in this volume. The desire is that the research provided here will be useful in the policy process in those many countries in the world which continue to view crop insurance as a possibly viable income stabilizing program.

This book is organized into three sections. Part one presents background material on crop insurance programs in the U.S., Canada and selected other countries. This section also includes some overall evaluations of these programs and suggests some reasons as to why their record of performance has not been good, when measured in terms of the excess of indemnity payments made over premiums collected, or participation, or as a possibly efficient substitute for direct income support programs. Part two provides some analytical models of multiple peril crop insurance which suggest the possibility of modification of design which could improve performance and which explores theoretical linkages between crop insurance decisions and other producer decisions previously not analyzed.

The main part of the book is part three where the results of a series of empirical studies using data bases particularly designed to answer crop insurance related questions are presented. This part of the book tests a number of the hypotheses which were raised in parts one and two regarding reasons for the widely held view by economists that crop insurance has not functioned well.

Part One—Historical Perspectives and Overall Performance

Part one begins with an overview of the fifty years experience of the United States with crop insurance by Gardner (chapter 2) who critically examines this experience in the context of other agricultural commodity programs. Gardner points out that crop insurance is related to other policies such as price supports, first because of its subsidy characteristic which influences expected profit, and secondly through its risk management characteristic which has an income stabilization effect. Since the stabilization effect has feedback in effects on expected prices and other variables, these two characteristics are, however, related. Crop insurance is found to have similar effects to some of the existing U.S. commodity programs but to work in opposition to other programs. For example, both crop insurance and the target price program increase output, but crop insurance works in opposition to the acreage diversion programs by expanding acreage. Gardner also points out that the increases in farming income stability caused by price support programs can be cushioned through the provision of subsidized crop insurance to provide income in years of crop failure.

Finally, Gardner argues that the data does not suggest that government disaster payment programs have reduced the demand for crop insurance as has been generally thought. He argues that one of the chief reasons this is true is that farmers have been allowed to receive disaster payments in addition to their crop insurance indemnities.

In his review of the U.S. experience with crop insurance, Gardner discusses a number of possible reasons for the low participation in multiple peril crop insurance programs during the 1980s which is viewed by many as the central "problem" with the program. Gardner points out that from 1981 to 1988 farmers paid out $1.6 billion in premium and received indemnity payments of $3.3 billion, for an average loss ratio of 2.1. Still, participation could not be pushed above 25 to 30 percent. He argues that the only hypothesis which is economically consistent with this data is that there is pervasive and large adverse selection and/or moral hazard problems with the program. These hypotheses are tested in part three of this book.

Finally Gardner discusses a number of options to reform the crop insurance program including its elimination which was recommended by the Bush administration and rejected by Congress in 1991. He points to a number of problems that would exist with all alternatives of which have been suggested. He is not optimistic about the possibilities of a successful reform of the existing system.

The thirty years' experience of Canada with multiple peril crop insurance has also been rather disappointing. Sigurdson and Sin, chapter 3, review the history of this program with a focus on its performance in the province of Saskatchewan. This chapter indicates that the crop insurance program in Canada has evolved over time to become more of an income support program, as measures such as increased premium subsidization were enacted to increase participation. These measures have been successful in that Canada currently has a participation rate of about sixty-five percent. The statistical analysis they present suggests that an increase in the expected rate of return to a farmer's purchase of crop insurance of about ten percent through premium subsidization can increase participation by about 1.85 percent. As might be expected, Canadian loss ratios, which average 2.83 from 1959 to 1988, are larger than that found in the current program of the United States.

Wright and Hewitt, chapter 4, find that loss ratios in general and experience worldwide with crop insurance has been uniformly negative. After providing an historical review of these experiences, in a number of countries, with a particularly interesting discussion of the rejection of crop insurance in Australia, Wright and Hewitt suggest that crop insurance participation is low because it may simply not be needed, partly because growers have other mechanisms for risk management available which have not been recognized by the theoretical literature.

Part Two—Conceptual Issues

In the first paper in this part, Quiggin, chapter 5, discusses a number of problems in the design of multiple peril crop insurance from a theoretical perspective. He points out that multiple peril crop insurance differs from other insurance problems in that there is little potential for pooling individual risks to obtain a less risky aggregate indemnity portfolio for the insurer, and that the problems of moral hazard and adverse selection can be more significant in an agricultural context.

Quiggin presents a conceptual alternatives to currently offered multiple peril crop insurance programs in a mechanism design framework. Quiggin considers the case where the optimal contract set consists of a range of deductibles and contingent contracts for various rainfall levels where the problem of moral hazard exists. If moral hazard problems are severe and prevent the emergence of contingent contracts based on yield levels, the optimal contract which emerges is a pure rainfall insurance contract. Public information on the state of the world such as rainfall (or probability distributions of such) is assumed to be available. Where this information is not available a standard multiple peril risk crop insurance will be optimal when it exists.

Under conditions of adverse selection, defined as heterogeneous farmer quality or land quality unknown to the insurer in advance, a separating equilibrium emerges as a possibility with the property that if the high risk groups are fully insured at actuarily fair rates the low risk group will only be partially insured. This equilibrium will be stable, of course, only if the insurer can prevent a high risk group from also purchasing contracts offered to the low risk group. In the cases of both moral hazard and adverse selection, the representation of optimal insurance in terms of contingent contracts yields new and interesting results.

Innez and Ardila in chapter 6 consider the problem of insurance from the viewpoint of interrelationship of production and investment decision under uncertainty. Innez and Ardila abstract from the standard problems of moral hazard, adverse selection, and distinctions between price and yield risk to focus on uncertainty arising from production and uncertainty from land values to examine the effects of government provided insurance on environmental quality related to production decisions such as the use of pesticides and soil depletion. The key result of this paper is that when land value risk dominates production risk, revenue stabilizing insurance can lead to increasing environmental externality from both sources of risk. Thus, this paper suggests that under conditions of instability with regard to land values, revenue stabilizing insurance may increase environmental problems; but governments may wish to explore the possibilities for land value insurance which can improve environmental quality. Some empirical evidence on the effect of multiple peril crop insurance in Canada on the use of marginal and submarginal lands can be found in chapter 13 of this book and is discussed briefly below.

Schmitz et al. in the final chapter of this part (chapter 7) analyze Canadian and United States crop insurance programs and give a detailed comparison of the interactions of multiple peril crop insurance with other agricultural commodity programs in these countries. It is noted that crop insurance in the state of Maryland is much less successful than in the northern great plains. Part of this is due to the limited possibilities for diversification in states like Montana. In Canada, crop insurance differs between provinces. The discussion in this chapter is concentrated on the prairie region. The analytical model suggests that the problems of moral hazard and adverse selection are greatly exacerbated by the possibilities of expected disaster payments and there is at least a theoretical reason for participation to decline in the presence of existing ad hoc disaster assistance programs.

Part Three—Applications and Policy Studies

Most of the empirical analyses of the impacts of multiple peril crop insurance in the United States contained in chapters 8 to 14 are based on data obtained from the Farm Cost and Return Survey (FCRS) which is conducted annually by the National Agricultural Statistical Service for the Economic Research Service. A brief description of this data base can be found in chapter 8 by Just and Calvin who also describe an extensive telephone survey which was designed to follow the FCRS to obtain additional information for the analysis of crop insurance participation. To our knowledge this is the first nationwide empirical analysis of microdata to examine the determinants of grower participation in the United States. Just and Calvin used a utility function with constant absolute of risk aversion to model the participation decision. This model allows them to test for the existence of adverse selection and moral hazard separately and the impact of these conditions on the participation decision. They find that heterogeneity of average yields among different producers of a crop is not an important source of adverse selection, but that heterogeneity of average yields across crops does suggest adverse selection and that federal crop insurance is largely being purchased by producers of high risk crops. They also find strong evidence of the existence of moral hazard associated with multiple peril crop insurance purchase decisions.

A probit analysis of participation decisions for corn supports the hypothesis advanced earlier in chapter 1 by Gardner that past experience with disaster payments tends not to have reduced participation in the crop insurance program. In fact, these regressions suggest that past experience with disasters that has a significant positive impact on the likelihood of insurance participation.

Additional evidence of the existence of adverse selection and moral hazard is presented by Quiggin et al. in chapter 9 who used the same FCRS data base to estimate an aggregate Cobb-Douglas production function using a dummy variable to discriminate between participants and non-participants in the program. In their model they do not differentiate between the problems of moral hazard and adverse selection, and give reasons why it is difficult to do so in an agricultural setting. They test the implications of the existence of these problems and find that indeed insured farmers have lower levels of variable input use and lower total factor productivity than uninsured farmers.

In chapter 10, Leathers, again using the FCRS data base explored the impact of financial characteristics of growers on crop insurance participation. His theoretical model suggests that farmers with low debt and high incomes are less likely to insure. Farmers with high debt and low net worth are more likely to be required to insure by lenders; and farmers with high debt and high net worth are more likely to insure voluntarily. The empirical analysis generally supports the analytical results, but it was not possible to construct a satisfactory statistical test of the hypotheses of insurance "imposed" by lenders. The regression parameter estimates, however, provide support for the hypothesis.

Schoney, et al. in chapter 11 examine the empirical relationship between crop selection and crop insurance and contrast the risk reducing effects of crop mix and insurance. Using representative farm data, it is determined that the risk-reducing effect of crop insurance is diminished under optimum crop selection, due to the portfolio diversification effect. The subsidization of premiums results in positive net payouts to the farmer on a crop basis for most crops. However, the small net premium on canola results in a net premium on a crop portfolio basis; the insured two-crop (wheat-canola) rotation cannot stochastically dominate the uninsured two-crop (wheat-canola) rotation; that is, canola tends to subsidize wheat in the premium structure. In areas which have potential to reduce risk through crop diversification, crop insurance acts primarily as catastrophe insurance.

Chapters 12 and 13 provide some empirical evidence on the impacts of multiple peril crop insurance on environmental quality in the United States and Canada, respectively. In chapter 12, Horowitz and Lichtenberg use the FCRS data base to examine the impact of crop insurance on pesticide and fertilizer use in corn production. The results suggest that crop insurance has a substantial influence on chemical use. Farmers who purchased crop insurance applied almost 20 pounds more nitrogen per acre and spent about $3.70 more per acre on pesticides. While crop insurance does tend to have some substitution effect for pesticides which are generally reviewed as risk reducing inputs, it is not significant. That crop insurance increased nitrogen usage was expected since nitrogen is generally regarded as a risk increasing input and thus both moral hazard and adverse selection would lead to this result. The positive impact of crop insurance on pesticides, although not as strong as nitrogen, was unexpected in that pesticides are usually regarded as risk reducing inputs. In fact, crop insurance premium subsidies have often been justified by their possible substitution effects for pesticides. Horowitz and Lichtenberg, however, show in this chapter that there are theoretical reasons why this result may be reasonable when there are multiple sources of uncertainty as there are in corn production.

Weisensel et al. find in their study that land use decisions do not change as a result of crop insurance in Saskatchewan. In the context of the theoretical model presented by Innez and Ardila in chapter 6, these results suggest the possibility of land value risk dominance over production risk if the model presented in chapter 6 can be broadly interpreted to include the possibility of increasing output through bringing submarginal lands into production.

To summarize these two chapters, there is evidence that crop insurance in the United States has had impacts on chemical use, primarily nitrogen, but land degradation at present cannot be attributed to crop insurance programs.

The final paper of this section, chapter 14, presents some simulation results of a modified deficiency payment program which would be based on the difference between the target revenue and average revenue in a region and the simultaneous elimination of target priced deficiency payments, crop insurance and ad hoc disaster assistance programs. The focus is on revenue stabilization through the exploitation of the natural negative correlation between price and yield. The simulations show that with the target revenue program producers can receive the same revenue as with deficiency payments based on target prices with the same government outlay, but significantly reduced variability of producer revenues and government expenditures. That is, under risk aversion there is a clear net benefit associated with this program. Since this program is a transparent pure income transfer rather than an insurance program, it would likely generate a negative response from producer groups and would be viewed as a direct subsidy in the context of the General Agreement on Tariffs and Trade negotiations. It is, nevertheless, an interesting proposal, and deserves further consideration by policymakers.

Conclusions

There is clear, effective, political demand for protection from climatic and market risks by agricultural producers in a great number of countries of the world. On the basis of the work found in this volume, and in an earlier book on this subject by Hazell et al. (1986), multiple peril crop insurance does not appear to be a promising alternative for supplying this protection. The existence of moral hazard and adverse selection problems associated with agricultural production have been solidly demonstrated in this book. It is doubtful whether anywhere near fair insurance premiums, using deductibles for the treatment of moral hazard, and separate policies for risk groups and for adverse selection with their appropriate enforcement, can be designed. Given the pervasiveness of moral hazard observed in the work presented here, however, the alternative of an institutionalized disaster payment program would be highly costly, and would also not be useful to growers at all as a risk management tool.

Revenue stabilization programs such as that suggested by Miranda and Glauber here are economically attractive but probably politically infeasible. One suspects that since crop insurance is not viewed as a subsidy internationally, or by growers, and because it can be viewed as a risk management tool, it is likely to continue to exist in most countries where it does now, and there will be increasing pressure and there will be increasing demand for it in other countries. Thus, efforts should continue to improve efficiency of the mechanism, but the focus on increasing participation levels is probably misplaced.

References

Hazell P, Pomareda C, Valdes A (1986) Crop insurance for agricultural development: issues and experience. Johns Hopkins Univ Press, Baltimore, London

Part I

Historical Perspectives and Overall Performance

Chapter 2

Crop Insurance in U.S. Farm Policy

B.L. GARDNER[1]

2.1 Introduction

Crop insurance is one of an interrelated set of governmental activities aimed at solving problems in the farm sector. The most commonly perceived problems are farm incomes that are (1) too low and (2) too unstable. This is not a normative judgment, but rather a statement about the political situation: it has not proved possible to reach political equilibrium without government intervention in aid of farmers. Nonetheless, taxpayer and consumer costs constrain the scale of such intervention. Thus, farm policy is essentially an attempt to support and stabilize farm income without imposing undue (i.e., politically unsustainable) burdens on taxpayers or consumers. Governmental provision of subsidized crop insurance is an element of such policies, and under certain circumstances has become a key element.

This chapter considers the role of crop insurance in the context of other components of farm policy, concentrating on the United States in the past two decades. The chapter consists of three sections: first, an outline of the scope and importance of crop insurance and other policies in recent years; second, a more detailed analysis of the U.S. crop insurance program in the 1980s; and third, the potential role of reformed and/or expanded crop insurance programs in future farm policies in the U.S. and abroad.

The first section indicates that crop insurance has generally been a small part of aggregate U.S. farm programs, accounting for less than 10 percent of U.S. commodity program budget outlays, and perhaps even less important in supporting farm income. But crop insurance has been very important in certain instances—wheat

---

[1]Department of Agricultural and Resource Economics, University of Maryland, College Park, MD 20742, USA.

in the High Plains, soybeans in the Southeast. And the achievement of substantial producer price stability has made farmers' losses because of crop failures more salient and increased the demand for crop insurance. But this demand has taken place in the political arena as a debate over disaster assistance, which has become the *de facto* dominant source of crop insurance.

The second part of the chapter reviews the U.S. crop insurance experience more specifically. It is not an inspiring story. Participation in the all-risk insurance program established by the Federal Crop Insurance Act of 1980 has remained stubbornly low, never exceeding 40 percent of eligible acreage. Consequently, Congress almost routinely has enacted disaster assistance programs which make payments analogous to insurance indemnity payments, but without charging premiums. About $1.5 billion annually was spent under disaster legislation in 1988, 1989, and 1991 (also covering 1990 disasters retroactively). At the same time the government's underwriting losses on its Federal Crop Insurance Corporation all-risk policies continued to increase, and averaged $300 million annually in the 1980s.

The third section considers alternative approaches to crop insurance in the future. Changes will be encouraged not only by past problems but also by the treatment of insurance measures in the Uruguay Round of negotiations in the General Agreement on Tariffs and Trade (GATT). Disaster assistance has come to be considered a relatively benign form of intervention in the GATT, and restraints on other farm income support policies will tend to induce the United States and other countries to redirect their policies in the direction of insurance programs. The options considered are: (1) reform of current crop insurance to improve actuarial soundness, (2) abandonment of current programs, shifting to full reliance on disaster programs, (3) mandatory insurance for program crops, (4) county yield insurance, and (5) expansion of crop insurance to cover more sources of income loss, providing revenue insurance that would replace current price support programs as well as crop insurance and disaster programs. Impetus for change is provided by the Omnibus Budget Reconciliation Act

(OBRA) of 1990 which applied meaningful disciplines to price support expenditures and to disaster relief expenditures. These constraints caused payments under the disaster legislation of 1991 to be prorated such that each eligible farmer received only 50 percent of the $2 billion in indemnity payments which the legislation authorized.

2.2 Interrelationship Between Crop Insurance and Farm Income Support

In political debate as well as economic theory, crop insurance is inextricable from other agricultural policies. The "pay as you go" feature of the budget agreement specified in the Omnibus Budget Reconciliation Act meant that any increased government spending on the Federal Crop Insurance Corporation (FCIC) program, or on disaster relief, had to be offset by reduced spending on other programs. Previously, under the Gramm-Rudman-Hollings law, supplemental appropriations for disasters were exempt from overall budget discipline. (This is still the case for Presidentially declared "dire emergencies," but the Bush Administration has so far not permitted this escape hatch to be used for agricultural crop failures.)

In the 1986-92 Uruguay Round of GATT negotiations, subsidies for crop insurance have been incorporated with other assistance in the "aggregate measure of support" (AMS) to be disciplined by member countries. The "Dunkel text" [GATT, 1991], put forward by the GATT as a draft agreement in agriculture, assimilated crop insurance in a broad category of "payments for relief from natural disasters," [p. L.16] itself part of a still broader aggregation of domestic agricultural support policies.

Analytically, the economics of public crop insurance fits in with other policies in two distinct but related ways: first, as a producer subsidy measure, in which respect crop insurance influences the expected values of commodity prices, farm incomes and other economic variables; and second, as a risk management measure. As a risk management tool, crop insurance provides stabilization services of value to farmers apart from any subsidies. Because these stabilization benefits will influence the mean values of prices and incomes, and because price supports increase the demand for quantity stabilization, the two aspects of crop insurance policy are interrelated.

Public assistance for income stabilization purposes can be justified on the grounds that informational asymmetries prevent development of a private insurance industry; so even unsubsidized public crop insurance would provide social benefits. However, if the public provision or subsidy of insurance cannot solve the problems which prevent private competitive insurance markets from being successful, then it is not likely that government intervention will increase welfare. For an analysis of factors that create insurability problems see Borch, 1990; Chambers, 1989; and Nelson and Loehman, 1987. This chapter does not consider the optimality of alternative crop insurance arrangements, but concentrates on the consequences of the policies that exist.

Some of the basic economics of crop insurance in the context of the market for an agricultural commodity can be seen in the case of U.S. grains. The main policy instruments of the 1980s are:

1. Market support prices or "loan rates" at which the Commodity Credit Corporation, an agency of the U.S. Government, grants loans and accepts delivery of commodities that maintain the market price at the support level.
2. Target prices, which determine the amount of a "deficiency" payment to farmers equal to the difference between the target price and the higher of loan rate or the U.S. average farm-level market price.
3. Acreage reduction percentages, a fraction of each farmer's historical acreage base of the crop which must be held out of production if the farmer is to be eligible for payments or CCC loans.

4. The Conservation Reserve Program which by 1991 has added 34 million acres of cropland, including about 15 percent of U.S. wheat acreage, under 10-year contracts, paying participating farms an average of $50 per acre annual rental rate.
5. Export enhancement payments, made in the form of title to CCC-owned stocks, to exporting firms equal to the difference between the U.S. price and the foreign price the firm establishes (to the satisfaction of the U.S. Department of Agriculture) as necessary to meet competition from other countries' exports.
6. Disaster payments, not legislated on a multi-year basis as items 1 to 4 are, but annually appropriated compensation for crop failure which has averaged about $1 billion per year ($600 million for grains) in 1985-91.

Crop insurance is a relatively small element of U.S. agricultural policy as measured by Federal budget outlays. Table 1 shows outlays for the main expenditure items. Crop insurance accounts for about 5 percent of all U.S. commodity program outlays. The percentage is smaller for the grains—some non-grain commodities, notably certain fruits and vegetables, get a substantially higher fraction of their support through crop insurance since these commodities do not have price supports.

Crop insurance may work either to reinforce or to contravene the effects of other policies. An example of reinforcement is that both crop insurance and the target price program tend to cause farm output to increase. But crop insurance will tend to contravene the output effects of the acreage diversion program. In recent years, loan rates have been kept too low to influence output appreciably, and target price and acreage reduction programs are estimated to have almost offset one another, so that the net effect on U.S. grain production is almost zero (for discussion see Gardner, 1990).

Crop insurance provides an incentive to greater output for two reasons: (1) the premium subsidy is effectively an input subsidy for farmers who buy insurance; and (2) the existence of

Table 1. Farm Program, Disaster, and Federal Crop Insurance Budget Levels, Fiscal Years

FY	(1) Farm Programs[a]	(2) Disaster Payments	(3) Federal Crop Insurance[c]
		$ Billions	
1981	9.8	0.3	0.2
1982	14.3	1.4	0.4
1983	21.3	0.3	0.5
1984	11.9	0.1	0.5
1985	23.8	0.0	0.5
1986	29.6	0.0	0.5
1987	24.7	0.0	0.3
1988	15.2	0.7	0.9
1989	14.8	3.9	0.7
1990	9.8	1.7[b]	0.6
1991	13.5	0.1	0.8
1992E	12.8	1.0	0.9

[a]Budget outlays under OMB function 351, "Farm Income Stabilization." Includes payments and administrative costs for all commodity programs. Excludes research and conservation activities (notably the Cropland Reserve Program). Source: U.S. Executive Office of the President, 1992.

[b]Includes $1.5 billion paid in kind using CCC commodities.

[c]Includes administrative costs ($297 million in FY 1991).

all-risk insurance encourages risk-averse producers to grow grains who otherwise might not.

To estimate the magnitude of the first effect, the premium subsidy plus pro-rata administrative costs averaged $200 million annually in 1985-90 for the grains. An average of 17 percent of the crop or roughly 2.0 billion bushels of grain was grown by participating farmers. The crop insurance subsidy thus amounts to about 10 cents per bushel or an average of about 3 percent of the price of grain for this 17 percent of output (and zero subsidy on the 87 percent of grain not covered by crop insurance). With an elasticity of supply of 0.3, the subsidy would increase U.S. grain production by a maximum of $.03 \times .3 \times .17 = .15$ of 1 percent (a maximum because this calculation assumes the market price is unaffected by the added output). This amounts to perhaps 18 million bushels (0.5 million tonnes) of grain. This is a small amount, but perhaps not completely negligible. The effect on output would be completely offset by releasing 0.5 million acres of the 11 million acres of wheat (base) land in the Conservation Reserve Program.

An example of larger effects is the production incentives provided by the Disaster Payments Program of 1974-81 on wheat acreage in marginal production areas of the U.S. High Plains. Censuses of Agriculture conducted in 1974 and 1978 provide evidence at the county level of the consequences of introducing an essentially free all-risk insurance program. Using data from Gardner and Kramer (1986), twelve marginal cropland counties are identified as those with a crop loss record so severe that FCIC insurance coverage was withdrawn from the county.[2] In these counties, disaster payments in 1976-79 averaged \$4.40 per acre, as compared to \$.90 per acre for the United States as a whole. Moreover, the disaster payments constitute a relatively high percentage of the rental value of land in these counties because the land is less valuable than U.S. average cropland.

Gardner and Kramer estimate that the payments increased the expected rental value of land in the marginal counties by about 12-15 percent. And indeed between 1974 and 1978 cropland

[2]The counties are: Texas: Armstrong, Andrews, Borden, Donley, Kent, Motley, Scurry, and Wheeler; Colorado: Kiowa; Georgia: Screven, Telfair, and Washington.

harvested in these counties increased by 20 percent, compared to 6 percent for all U.S. counties.

Although the evidence has not been analyzed systematically, it is also likely that soybean acreage in the Southeast is larger than it would be in the absence of subsidized all-risk insurance. Generally, the subsidy levels vary widely by crop and region, so the insurance raises the returns to growing some crops significantly more than the calculations for grains indicate. Table 2 presents information by commodity for 1988-90. Peaches, potatoes, and rice are crops which seem able to count regularly on receiving indemnities exceeding insurance premiums, while grapes regularly sow more than they reap. Corn is the one crop whose experience generates data that look like the long-term expectations for an actuarially sound insurance industry—average indemnities are close to premiums—although even here the insurer takes a loss. And the participation rate for corn is only about 10 percent.

In recent years, the major crop insurance issue has been its interrelationship with Federal disaster legislation. The standing Disaster Payments Program expired with the enactment of the Agriculture and Food Act of 1981. But annual disaster legislation effectively replaced this program, albeit on a less predictable basis. Major disaster relief bills were enacted in 1988, 1989, and 1991. The amounts paid out are shown in Table 1.

It is argued that the existence of disaster relief reduces the demand for crop insurance. For example, the President's Budget for FY 1992 states:

> Federal Crop Insurance illustrates the difficulty in selling disaster insurance, even when heavily subsidized. Federal crop insurance was expanded in 1980 to protect farmers from the risk of crop loss and obviate the need for disaster relief. The program is in need of serious reform.
>
> Only one-third of eligible commodity production has been insured, despite the fact that subsidies cover about two-thirds of total program

Table 2. FCIC Insurance by Commodity, 1988-90

Crop	Year	Premiums (million $)	Indemnities (million $)	Loss Ratio	Three-year Ratio
Apples	1988	5.7	12.8	2.2	
	1989	7.7	13.7	1.8	2.2
	1990	5.9	16.2	2.7	
Barley	1988	11.2	52.9	4.7	
	1989	18.0	26.5	1.5	2.0
	1990	19.9	18.2	.94	
Corn	1988	86.3	274.6	3.2	
	1989	267.3	243.9	.91	1.1
	1990	213.9	115.9	.54	
Cotton	1988	39.9	36.4	.91	
	1989	52.8	126.8	2.40	1.5
	1990	81.0	102.5	1.26	
Grapes	1988	2.4	0.5	.21	
	1989	3.1	2.1	.68	0.6
	1990	3.5	2.5	.71	
Peaches	1988	1.5	3.5	2.33	
	1989	1.7	7.9	5.26	3.7
	1990	2.0	8.0	4.00	
Peanuts	1988	20.3	31.7	1.56	
	1989	26.1	33.3	1.28	3.5
	1990	30.0	198.8	6.63	
Potatoes	1988	6.2	17.9	2.89	
	1989	8.2	20.3	2.48	2.6
	1990	10.7	27.1	2.53	
Rice	1988	2.7	7.4	2.74	
	1989	2.7	7.9	2.93	3.1
	1990	3.4	11.7	3.44	
Wheat	1988	69.1	260.2	3.77	
	1989	138.8	316.5	2.28	1.9
	1990	199.2	178.3	.90	

Source: USDA, 1991, pp. 401-403.

> costs, including premiums, administration, and excess losses. Yet even with a subsidy of roughly \$1 billion per year, disaster relief bills costing \$7 billion were enacted in 1983, 1986, 1988, and 1989. (U.S. Executive Office of the President, 1991, part II, p. 216.)

Goodwin (1991), Hojjati and Bockstael (1988), Calvin (1990), Niewoudt et al. (1985), and Barnett et al. (1990) have all found that participation in crop insurance varies significantly with the ratio of indemnities to premiums. Yet this evidence is not sufficient to nail down the effect of disaster payments on crop insurance demand. The reason is that buyers of crop insurance have been allowed to receive disaster payments <u>in addition to</u> their crop insurance indemnities. So the expected returns from buying crop insurance have not been reduced by the existence of disaster programs. Nonetheless, to the extent that farmers buy crop insurance because of risk aversion (rather than expected returns), the existence of reasonably reliable disaster assistance must reduce the demand for crop insurance.

A related argument is that subsidized crop insurance and disaster programs encourage farmers not only to grow more of the covered crops, but also to grow crops in more risky circumstances—with less diversification, less use of precautionary measures such as fallowing, stand-by irrigation, and pesticide applications. Gardner and Kramer (1983) attempted to test for these effects of the 1970s disaster program, but did not find significant effects in the marginal counties studied.

Commodity price supports and subsidized crop insurance are interrelated in two further ways. First, output increases in response to subsidized insurance add to surplus production and hence the cost of farm programs. Second, the existence of price supports has the effect of increasing the instability of farmers' incomes caused by random output fluctuations, particularly widespread crop failure. This issue received its classic analysis in McKinnon (1967). Its application to supported commodities can be simply expressed:

with a producer price floor near the cost of production, the only way a farmer is exposed to significant losses is through crop failure. Moreover, governmental stabilization of year-to-year fluctuations in market prices, through storage and varying set-asides, means that high market prices will not cushion the blow of widespread crop failure as much as would be the case without the stabilization programs.

These considerations have important implications for future crop insurance policy options, but before exploring this subject, the next section considers the U.S. experience in crop insurance in more detail.

2.3 The U.S. Experience in Crop Insurance

The U.S. government has administered a crop insurance program since 1938, when legislation created the Federal Crop Insurance Corporation (FCIC). The FCIC is an agency of the U.S. Department of Agriculture, with no privately held stock and management composed of government officials. The FCIC program began modestly, insuring only wheat in 1939. But from the start the program confronted two problems that persist to this day: low participation by farmers and an excess of indemnities paid out as compared to premiums paid in. The "loss ratio," indemnities paid by the FCIC to cover crop losses divided by premiums received from farmers (including premium subsidies paid by the government) averaged 1.54 in 1939-41. During 1980-88 the loss ratio averaged 1.51.

An underlying reason for the low participation rate and high loss ratio is adverse selection which occurs when the insurer, lacking information on individual farmers' situations, treats each producer equally and as if the current year were an average year. But the farmers themselves have a better notion of their own situation, and will decide whether to buy insurance with that information in mind. Adverse selection plagued the FCIC from the beginning. When wheat-yield insurance was selling slowly in the fall of 1938, the sign-up period was extended to March 1939 (and farmers were permitted to use advance Agricultural Adjustment Act payments to cover the premium). But by the spring of 1939, lack of soil moisture in many areas made it quite clear that the chances of a crop failure were unusually high. Clendenin (1942) found generally that "In counties and years when soil moisture was lacking, insurance sold freely. In those same counties the number of contracts dropped as much as 75 percent when soil-moisture conditions presaged a good crop" (p. 249).

Continuing problems of underwriting losses led to the cessation of federal crop insurance, which had been offered only for wheat and cotton, in 1943. But FCIC programs were revived immediately in the post-War years, and have remained in place to the present. A summary of FCIC receipts from insurance sales and indemnities paid out in 1988 is shown in Table 3.

2.3.1 The 1980s Experience. The Federal Crop Insurance Act of 1980 was intended to make major improvements in crop insurance programs. While the average loss ratios for the 1960s and 1970s did not indicate financial problems with the FCIC, two developments in the 1970s created difficulties. First, the Agriculture and Consumer Protection Act of 1973 established a disaster payments program, which was renewed and modified in the Food and Agriculture Act of 1977. The disaster payments program essentially provided free insurance to producers of the major crops (wheat, rice, feed grains, and cotton), if they participated in price support programs. This insurance was comprehensive in its coverage of risk, in some cases even making "prevented planting" payments if adverse conditions prevented a farmer from getting a crop in the ground. Disaster payments totaled $3.4 billion in 1974-80, almost $500 million annually. Second, participation in FCIC programs remained low. Less than ten percent of potentially insurable acreage was enrolled. Aggregate FCIC indemnities in 1974-80 totaled $0.9 billion.

Table 3. FCIC Insurance Premiums Received, Indemnities Paid and Loss Ratio, 1939-88

Year	Premiums (million dollars)	Indemnities (million dollars)	Loss Ratio
1939	3.4	5.6	1.64
1940	9	14	1.51
1941	11	19	1.68
1942	17	25	1.49
1943	18	33	1.82
1944			
1945	9	23	2.48
1946	35	63	1.80
1947	44	35	.81
1948	13	7	.53
1949	12	16	1.31
1950	14	13	.91
1951	19	21	1.12
1952	21	21	.97
1953	27	31	1.15
1954	23	28	1.24
1955	22	26	1.14
1956	22	28	1.26
1957	17	12	.69
1958	18	5	.26
1959	18	14	.77
1960	18	10	.58
1961	18	16	.89
1962	22	24	1.10
1963	30	24	.77
1964	34	30	.90
1965	36	41	1.13
1966	37	25	.68
1967	43	55	1.27
1968	49	51	1.05
1969	49	53	1.08
1970	44	42	.94
1971	48	29	.60
1972	42	25	.60
1973	48	28	.60
1974	54	63	1.17
1975	73	63	.86
1976	91	142	1.57
1977	102	149	1.46
1978	94	47	.50
1979	103	65	.65
1980	157	359	2.28
1981	364	407	1.12
1982	596	517	.86
1983	285	580	2.04
1984	429	619	1.44
1985	438	653	1.49
1986	379	612	1.61
1987	364	366	1.01
1988	394	954	2.42
1989	819	1208	1.47
1990	836	1030	1.23
1991E	736	956	1.30

Source: Federal Crop Insurance Corporation

It might be thought that the low participation in FCIC programs was attributable to the disaster payment program, but substitution is not apparent in the data. In the four years before the disaster payment program, 1970-73, farmers paid $46 million annually for FCIC insurance. During 1974-80, farmers' premium payments were $96 million annually. Nonetheless, it became clear that with free disaster protection available, it would be difficult if not impossible to expand FCIC participation beyond the 10 percent level of 1980.[3] By the end of the 1970s, even the Secretary of Agriculture, Bob Bergland, was saying that the disaster program had become a disaster itself.

In order to place the FCIC on a sound business footing and replace the disaster payments program, the following steps were taken under the Federal Crop Insurance Act of 1980: (1) authorization of up to a 30 percent premium subsidy (in addition to governmental absorption of administrative costs); (2) expansion of coverage to new counties and new commodities; (3) authorization of commission payments to private companies for FCIC sales and service; and (4) reduction of FCIC premiums for producers who bought commercial hail or fire insurance. These steps, however, proved insufficient to place the FCIC on a sound business footing or to forestall Congressional issuance of *ex post* free insurance through disaster bills.

The experience under the 1980 Act has been disappointing in three main respects. First, while participation increased from 10 percent of potentially insurable acreage in 1980 to 25 percent in 1988, the program's goal of 50 percent participation was not reached (Table 4). Gardner and Kramer (1986) estimated that a county's participation rate increased by about 0.3 percentage points with each 1 percent increase in the rate of return from buying

---

[3]The 10 percent participation rate is measured as acreage insured as a fraction of acreage potentially insurable, using the "net" acreage concept (adjusted for crop-sharing contracts in which only one party insures) developed by the Commission for the Improvement of the Federal Crop Insurance Program. For a discussion of difficulties in obtaining a meaningful measure of the participation rate, see the U.S. Commission on Crop Insurance (1989, pp. 41-56).

Table 4. FCIC insured acres[a] as a percentage of planted acres of insurable crops

Year	Percent
1980	9.6
1981	15.9
1982	15.3
1983	11.6
1984	15.5
1985	18.2
1986	19.6
1987	21.9
1988	24.5
1989	40[b]
1990	40[b]
1991	32[c]

[a]The concept of "net insured" acres is used (see U.S. Commission for the Improvement of the Federal Crop Insurance Program, 1989, p. 41). This means that under a 50-50 share contract, if the tenant buys insurance while the landlord does not, each 100 acres counts as only 50 insured acres.

[b]The 1989 and 1990 figures do not attest to an increase in attractiveness of the FCIC program because producers who received drought assistance in 1988 were required to buy FCIC insurance in 1989 as a condition of eligibility for price support programs.

[c]FCIC estimate.

insurance. Assuming that the 30 percent premium subsidy increases the expected rate of return by 30 percent, the participation rate should be increased to about 20 percent under the 1980 Act. The more sophisticated insurance demand model of Hojjati and Bockstael (1988) implies that a 1 percent premium subsidy should increase participation by 0.5 percentage points. According to this estimate, the 30 percent premium subsidy should have increased participation to about 25 percent in the 1980s. The actual participation rate in 1981-88 averaged 18 percent, but ended the period (not counting the mandatory insurance policy of 1989 and 1990) at a 32 percent rate. Thus, participation has been roughly consistent with models that explain insurance purchases by the price (premium) charged as compared to expected indemnities received. The greater sales effort that the 1980 Act was supposed to have unleashed does not appear to have had a large effect.

The second problem with the 1980 Act is that the loss ratio of indemnities paid out to premiums paid in to the FCIC (including government premium subsidies) increased to 1.41 in 1981-89, compared to 1.10 in 1948-80 (U.S. Commission for the Improvement of the Federal Crop Insurance Program, 1989, p. 234). The budgetary costs totaled $3.7 billion or an average of $465 million annually in 1981-88. This is much more than the $85 million annual premium subsidy that the 30 percent subsidy rate generated. Table 5 provides details of the government's costs. The actual rate of return to buying insurance can be roughly estimated from the $200 million in premiums paid by farmers on average in 1981-88 compared to the $400 million actually paid out in indemnities.

Third, while the disaster payments program was phased out in the 1980s, a new series of disaster bills was enacted in low-yield years which cost the government, on average, even more than the disaster payments program had cost. Spending for ad hoc disaster bills averaged $600 million annually in 1981-88, even larger than the $500 million annual average cost of disaster payments in 1974-80.

The high costs of the 1980 Act are undoubtedly attributable in part to a lack of actuarial and management experience in new crops, new counties, and new insurance marketing approaches. But there is no trend toward improvement over time in the data shown in Table 5, even excluding the 1988 drought year. Examination of loss ratios by commodity indicates that problems were as great for traditionally insured commodities as for commodities newly insured

Table 5. Government costs of FCIC insurance

Year	Premium subsidy	FCIC adminis-tration	Commissions to private marketers	Reinsurance costs	Excess of indemnities over premium	Total
			$ million			
1981	47	61	28	4	30	169
1982	92	69	47	23	132	364
1983	64	70	26	36	398	493
1984	98	74	25	79	204	480
1985	100	79	18	103	242	542
1986	88	85	11	98	234	515
1987	88	60	13	97	5	262
1988	108	65	12	122	586	891
1990	215	n.a.	n.a.	n.a.	196	n.a.
1991(est.)	170	n.a.	n.a.	n.a.	220	n.a.
Average (1981-88)	86	70	22	70	216	465

n.a.: not available

Sources: U.S. Commission for the Improvement of the Federal Crop Insurance Program, 1989, Table 5, p. 235 (1981-88). Federal Crop Insurance Corporation, 1990-91.

under the 1980 Act. Moreover, with the exception of cotton and citrus the loss ratios for the traditionally insured crops increased under the 1980 Act (see Table 6).

Overall, there are clear reasons why the President's Budget for fiscal year 1991 proposed that governmentally underwritten crop insurance be abandoned. The Administration's proposal for the 1990 farm bill was the following recommendation (U.S. Dept. of Agriculture, 1990, p. 65):

Table 6. Loss ratios by crop

Insured prior to 1981	1948-80 average	1981-88 average
Wheat	1.05	1.71
Cotton	1.57	1.32
Tobacco	1.04	1.24
Corn	1.08	1.15
Grain Sorghum	1.05	1.62
Barley	0.95	2.41
Oats	0.84	1.81
Soybeans	0.89	1.90
Peanuts	1.35	1.64
Sugar Beets	0.77	1.11
Sugar Cane	0.86	1.73
Forage	2.00	2.53
Citrus	1.40	1.39
Potatoes	2.06	2.76
Apples	1.27	1.94
Tomatoes	0.75	0.78
Newly insured under 1980 Act		
Onions		2.38
Grapes		1.60
Cranberries		0.81
Almonds		1.34
Peppers		1.23
Popcorn		2.14
Walnuts		0.84

Source: U.S. Commission for the Improvement of the Federal Crop Insurance Program, 1989, p. 274.

> Repeal legislation for the Federal Crop Insurance Program and propose legislation that would establish a standing disaster assistance program administered

through ASCS [Agricultural Stabilization and Conservation Service] that would pay producers in the event of an area-wide catastrophic loss:

• Eligibility would include those crops currently covered by FCIC, plus hay and forage. These crops account for over 93 percent of the total U.S. cropland.

• Disaster payments would be available for producers of crops in counties where actual county average harvested yield falls below 65 percent of normal as estimated by the National Agricultural Statistics Service (NASS). If data are insufficient at the county level to ensure reasonable assessments of normal harvested yield, regional or State yields will be used.

• Once a county is declared eligible, individual farmers receive disaster payments for any shortfall in their own farm harvested yield below 60 percent of the NASS county average yield for harvested acres.

• Payments for reduced yield and failed acreage would be based on 65 percent of the 3 year average market price for each eligible commodity. Payments for prevented planting would be based on 33 percent of the 3-year average market price.

• Producers would be subject to a $100,000 payment limitation and would not qualify for payments if their gross annual income exceeded $2 million, similar to recent disaster legislation. Producers would be required to comply with conservation provisions of the Food Security Act of 1985.

• Producers receiving disaster payments under this program would not be eligible for FmHA Emergency Loans or CCC emergency feed program for the same losses.

Congress rejected this proposal; but the Food, Agriculture, Conservation and Trade Act of 1990, and the President's Budget for fiscal years 1991 and 1992, made it clear that the existing multi-peril crop insurance system is seen by both the Legislative and Executive Branches as being in severe need of overhaul.

The 1990 Farm Bill made only minor changes in crop insurance, but the Conference Committee Report of the House and Senate stated (U.S. Congress, 1990, p. 1183):

> The Managers intend that the crop insurance provisions in this Act do not represent an answer to the problems facing Federal crop insurance. A more fundamental restructuring of the existing program is needed to prevent the continued financial losses, low participation rates, and other inefficiencies that have plagued this program and required the enactment of repeated ad hoc disaster bills during the 1980s. Congress must address this problem soon, and the House and Senate Committees on Agriculture intend to revisit this issue in the 102nd Congress.

2.4 Lessons for Reform of Crop Insurance

Congressional and Executive Branch dissatisfaction with the current situation practically ensures substantial changes in crop insurance. But what the changes will be and when they will occur is not clear. It is now (June 1992) clear that the 102nd Congress will not revisit the issue in any serious way.

The main reason for hesitation is that there are substantial problems with all the available options. Nonetheless, the 1980s experience should help in weighing up the pros and cons of these options.

1. Further reform of FCIC programs. This approach would continue the path of the 1980 Act, but with further changes to correct problems that became apparent in the 1980s. The main

drawback to this approach, which at first glance appears insuperable, is that the FCIC programs of the 1980s were apparently profit-generating to the average farmers who bought insurance. In 1981-88, farmers paid about $1.6 billion in premiums to the FCIC and received indemnity payments of about $3.3 billion. Yet only 25 to 30 percent of eligible acreage could be induced into the program. What conceivable reform, short of giving FCIC policies away, could generate a significant increase in participation?

It is possible, however, that adverse selection is keeping participation low. The acreage in the program is, under this hypothesis, especially risky acreage, and moreover with risks not reflected fully in the premiums charged. Therefore, buying insurance is profitable for these farmers, as the data reflect. However, the nonparticipants, according to this view, have a very different situation. Their risks are low, but FCIC premiums, even with the 30 percent subsidy, are not low enough to make crop insurance a reasonable purchase even for a moderately risk-averse producer.

A potential solution to this problem would be to reduce FCIC premiums on less risky production. A key question, however, is whether adverse selection precludes this reform because the farmer knows the riskiness of production but the insurer does not? This asymmetry of information means the insurer does not know which acreage is more risky and so has to charge the same premium to all.

It seems evident that FCIC insurance is not confronted with a pure adverse selection situation. A crude disaggregation by state shows wide variation in both loss ratios and participation rates. For example, Montana, North Dakota, and Washington insured respectively, 68, 52, and 45 percent of all planted acres in 1988, while Ohio, Wisconsin, and Tennessee insured 6, 4, and 3 percent (U.S. Commission for the Improvement of the Federal Crop Insurance Program, 1989, p. 215). Table 6 shows longstanding variations in loss ratio by commodity. Therefore, participation could perhaps be increased without large underwriting losses by reducing premiums in low-participation states or commodities.

Moreover, although there still may be a significant adverse selection problem within the set of producers of a given commodity in a given state, even this could be overcome in time by having the FCIC experience-rate each farm (as automobile insurers do drivers).

Nonetheless, it is doubtful that the necessary reforms could be accomplished quickly or without a lot of exploratory work. Until the 1980 Act, the FCIC program as a whole was regarded as experimental, with legislated limits on expansion to new counties and commodities. The fact that FCIC efforts to date have not resulted in a program that draws high farmer participation at reasonably low government cost has to count heavily against the prospects of success with further reform.

2. Abandonment of the FCIC. This approach could be seen as a move to laissez faire in insurance—which already exists for hail and fire insurance. In this situation, it seems certain from past political results that Congress will enact disaster relief bills when crop failure afflicts any substantial geographic area. But even if these bills cost an average of $800 million annually, which is consistent with recent Congressional behavior, the total governmental cost could well be less than under the current set of programs. On the other hand, the absence of FCIC programs might lead to still larger and more frequent disaster bills than we have seen in the 1980s.

A more general objection to the disaster bill approach is that it has few of the desirable features of FCIC insurance and adds new undesirable features. With FCIC insurance a farmer can undertake genuine risk management—projects can be undertaken with high expected return but a distinct possibility of a ruinous outcome if the weather, say, goes wrong. The disaster bill approach does not accommodate individual risk management and in this sense is not insurance at all.

Moreover, the moral hazard problem, in which the insured party changes actions in ways that increase the likelihood of crop failure, is not overcome and may be a worse problem under disaster bills. With FCIC insurance the farmer at least had to pay a premium that covered a substantial part of the expected loss, and had in principle to adhere to good farming practices. Under disaster bills, the farmer is covered even if a field is planted on a flood plain or in a droughty area where crops fail often. The problem is analogous to building vacation houses on vulnerable beach lots under the protection of federal disaster programs.

Finally, because farmers pay part of the premium under FCIC insurance, it is not clear that there will be a sustained budgetary saving under the disaster bill approach.

3. Mandatory insurance for program crops. If it is accepted that feasible reform of FCIC insurance is not likely, for the short term at least, and that reliance on disaster bills is unacceptable on the grounds of inefficiency, high budget costs, too little risk management, and too much politics in determining payments, then a search for a third approach is necessary. Requiring participants in price support programs and FmHA loan programs to purchase crop insurance is a possibility. A premium subsidy could be foregone since there are already subsidies incorporated in the commodity and loan programs. Participation would be high given that price-support participation is about 80 percent for the major crops. Soybeans presents a problem, but few farmers grow only soybeans. With high participation, the pressure for disaster relief bills in response to area-wide crop failures would be lessened since most producers whose crops fail would receive indemnity payments under the crop insurance program.

Substantial farmer resistance could be anticipated if the required insurance consisted of FCIC policies as currently administered. Too many Corn Belt farmers find these contracts of negligible value, so that the required FCIC premium would be viewed simply as a tax. To avoid this problem an area-based insurance program might be substituted for FCIC contracts.

4. County Yield Insurance. This option would trigger an indemnity payment for every farmer in a county if the county-average yield fell below 90 percent, say, of its trend level. Each farmer would receive the county's bushel-per-acre shortfall times his farm's program acreage. The general approach goes back to the proposal of Halcrow (1949). Because a farmer can do very

little to influence the existence or size of the payments received, moral hazard problems are eliminated. Adverse selection still exists to the extent that farmers can opt in or out of the program in response to planting season weather. However, when crop insurance is least valuable, that is with favorable crop prospects and hence lower than average prices, price support protection will be particularly valuable and hence farmers will tend to buy crop insurance to maintain their eligibility for commodity price support programs. Because insurance indemnity payments are triggered by the same conditions that would otherwise trigger disaster-bill assistance, ad hoc disaster legislation should be effectively forestalled. And with no premium subsidy, the budgetary costs would be small.

The drawback of county-yield insurance is that it does not provide a risk management tool for the individual farmer. Several studies (see Miranda, 1991 and Williams et al., 1991) have attempted to quantify and evaluate the exposure to risk from a farmer having a disastrously low yield while the county as a whole has yields that are relatively normal. The results have not been definitive, but are encouraging enough that the authors have recommended further investigation. In 1992 pilot projects are under way to try out area-wide insurance coverage in soybeans and wheat.

5. Joint crop/price insurance. Twenty-five years ago McKinnon (1967) laid out clearly the symbiotic relationship between stabilization of price risk and output risk. In the practical policy vein, both price supports and crop insurance have been addressed in the Food, Agriculture, Conservation and Trade Act of 1990. Administratively, disaster payment programs have been brought to farmers by the Agricultural Stabilization and Conservation Service (ASCS), which also administers farm commodity programs. The President's fiscal year 1993 budget requests for the first time explicit appropriations for the administration of some Federal Crop Insurance Corporation program activities through ASCS offices. In short, both theory and

practice are moving toward a unified program of crop insurance and price support.

A possible end-point of this unification is a revenue insurance scheme. Canada has already introduced the approach. In the U.S. context, it would be fairly straightforward to base deficiency payments not on the shortfall of market price from the target price, but instead on the shortfall of revenue (market price times actual output) from the target price times normal output.

The "price insurance" provided by deficiency payments works from the U.S. average market price rather than the individual farmer's price. The corresponding revenue insurance approach would be to use area yield rather than individual yield as the basis for a farmer's output in the revenue shortfall calculation. Glauber and Miranda (this book) simulate the operation of such a "target revenue" program for the U.S. corn market. Their results indicate that the target revenue approach is more effective at stabilizing farm income than separate target price and crop insurance programs.

Since farmers now pay a premium for crop insurance, the revenue insurance approach suggests a premium for the price-insurance component also. This would provide a method of budgetary savings more flexible and beneficial to farmers than the relatively crude budgetary hatchet wielded by the current 15 percent non-payment base for deficiency payments.

Finally, as mentioned earlier, an agricultural agreement in the Uruguay Round of GATT negotiations is likely to provide impetus to insurance schemes generally, on the grounds that they tend to be less trade-distorting ways of supporting agriculture than other methods of farm support.

2.5 Summary

The Federal Crop Insurance Act of 1980 was intended to make major improvements in the crop insurance program which had been in operation since 1938 with mixed results. The immediate impetus for reform was the Disaster Payments Program of the Agriculture and Consumer Protection Act of 1973. The Disaster Payments Program provided free insurance to producers of the major crops. This insurance was comprehensive in coverage, in some cases even making prevented planting payments. These payments totaled $3.4 billion from 1974 to 1980, or almost half a billion dollars annually. In addition, participation in FCIC programs remained very low, with less than 10 percent of potentially insurable acreage enrolled.

The experience under the 1980 Act has in turn been disappointing. There are three main problems. First, while participation increased from 10 percent to 32 percent, the program's goal of 50 percent participation was never approached. Second, the loss ratio of indemnities paid out to premiums paid in to FCIC increased to 1.41 from 1981 to 1989 compared to 1.10 from 1948 to 1980.[4] The budgetary costs totaled $6.8 billion, or an average of $570 million annually from 1981 to 1992. The third problem is that despite these costs, participation has been too low to forestall a series of disaster bills, which cost the government on average even more than the Disaster Payments Program of the 1970s had cost. Program spending for disaster bills averaged about $600 million annually in the 1980s.

The result is a situation in which neither the current crop insurance program nor any easily specifiable alternative looks very attractive. In the 1990 farm bill debate, the Administration proposed a standing disaster assistance program that would provide area-based payments to producers which would cost approximately one-third as much as the current disaster bills would cost. But Congress rejected this approach, calling for administrative reforms of current programs, and further study. Neither the reforms nor further study have achieved significant results as of 1992, but some promising ideas have been explored—area-yield insurance,

---

[4]Even with reforms being implemented, the President's Budget contemplates loss ratios of about 1.3 through 1993.

experience rating, revenue insurance. Evolution of U.S. policy in one or more of these directions is highly likely.

References

Barnett BJ, Skees JR, Hourigan JD (1990) Examining participation in federal crop insurance. Staff Paper No 275, Dept Agric Econ, Univ Kentucky

Borch KH (1990) Economics of insurance. North-Holland, New York

Calvin L (1990) Participation in federal crop insurance. Pap presented S Agric Econ Assn Annu Meet, Little Rock

Chambers RG (1989) Insurability and moral hazard in agricultural insurance markets. Am Jour Agr Econ 71:604-16

Clendenin JC (1942) Federal crop insurance in operation. In: Wheat studies of the food research institute Vol XVIII, No 6, pp 229-290

Gardner B (1990) The why, how and consequences of agricultural policies: the United States. In: Sanderson FHS (ed) Agricultural protectionism. Resources for the Future, Washington, DC, pp. 19-63

Gardner B, Kramer R (1986) Experience with crop insurance programs in the United States. In: Hazell P and others (eds) Crop insurance for agricultural development. Johns Hopkins Univ Press, Baltimore

General Agreement on Tariffs and Trade (1991) Draft agreement on agriculture. MTN.TNC/W/FA

Glauber JW, Miranda MJ (this book) Providing catastrophic yield protection through a target revenue program

Goodwin BK (1991) An empirical analysis of the demand for multiple peril crop insurance. Mimeo, Dept Agric Econ, Kansas State Univ

Halcrow HG (1949) Actuarial structures for crop insurance. J Farm Econ 21:418-43

Hojjati B, Bockstael N (1988) Modeling the demand for multiple peril crop insurance. In: Mapp H (ed) Multiple peril crop insurance, So Cooperative Series Bul No 334 pp 153-176

McKinnon RJ (1967) Futures markets, buffer stocks, and income stability for primary producers. J Polit Econ 75:844-61

Miranda MJ (1991) Area-yield crop insurance reconsidered. Amer J Agr Econ 73:233-242

Nelson CH, Loehman ET (1987) Further toward a theory of agricultural insurance. Am Jour Agr Econ 69:523-31

Niewoudt WL, Johnson SR, Womack AW, Bullock JB (1985) The demand for crop insurance. Agricultural Economics Report No 1985-16, Dept Agric Econ, Univ Missouri

US Commission for the Improvement of the Federal Crop Insurance Program (1989) Recommendations and findings to improve the federal crop insurance program

US Congress (1990) Conference report to accompany S 2830. House of representatives Report 101-916

US Department of Agriculture (1990) The 1990 farm bill: proposal of the administration. Washington DC

US Executive Office of the President (1991) Budget of the US government: fiscal year 1992. Part 7, Table 3.2

Williams JR, Carriker GL, Barnaby GA, Harper JK, Black JR (1991) Area-measured crop insurance and disaster aid for wheat and grain sorghum. Mimeo, Dept Agr Econ, Kansas State Univ

Chapter 3

An Aggregate Analysis of Canadian Crop Insurance Policy

D. SIGURDSON[1] and R. SIN[2]

3.1 Introduction

In 1959, the federal government of Canada passed the Canadian Crop Insurance Act thus allowing provincial governments to establish crop insurance programs with financial support from the federal government. In this design it was recognized that the federal government had no constitutional authority to assist agriculture and that provincial governments did not have the resources to support a crop insurance program. It is important to understand both the events that led to the introduction of crop insurance in Canada and the programs which preceded the plan.

Government participation in crop insurance has influenced producer involvement in the program. This paper examines both the level and impact of government involvement. The impact of government participation is examined both from a distributional aspect and in regard to the impact on producer enrollment in the program.

3.2 Historical Review of Crop Insurance

One of the most disastrous periods in Canadian agriculture was the 1930s. In 1937, due to prolonged drought, the yield of wheat in Saskatchewan was 2 bushels per acre. Also during this period

---

[1]Treasury Board Branch, Saskatchewan Department of Finance, Regina, Saskatchewan, Canada.

[2]Economic and Fiscal Policy Branch, Saskatchewan Department of Finance, Regina, Saskatchewan, Canada.

there were no programs in place that farmers could use to protect or stabilize their incomes. The result of the devastating low crop production and a lack of income protection programs was a poverty-stricken agriculture sector and a mass exodus from the agriculture industry.

3.3 Prairie Farm Assistance Act

One of the first programs in Canada to offer income protection to producers resembled a crop insurance program. This program was the Prairie Farm Assistance Act (PFAA) which was introduced in 1939. The objectives of PFAA were never clearly stated. It was, however, to act as an acreage insurance plan as well as providing relief support to farmers.

The PFAA was implemented by the federal government in response to the drought which was destroying the prairie agricultural industry and economy. Farm population was declining, and it was felt that some form of assistance was needed to protect the industry and the participants in that industry. The government also felt partially responsible for the plight of farm families in that they had encouraged farming in the prairies as a means of creating a market for Eastern Canadian goods and thus fostering the country's economy. Originally the settlement was successful as the prairies prospered, with agriculture the mainstay of the economy. However with the drought, beginning in 1930, questions were raised about the viability of the area. This is reflected in the debate in Parliament when the Minister of Agriculture stated (Standing Committee on Agriculture and Colonization):[3] "... P.F.A.A. is intended to take care of people who were put on land that they should not have been put on. That is our reason for being in this at all and it is our reason for paying two-thirds or three-quarters of the cost out of the treasury of Canada. We say that

[3] Minutes of Proceedings and Evidence, Standing Committee on Agriculture and Colonization, House of Commons, Ottawa, June 10, 1947, p. 46.

these people, with our consent and with our help, were put on land which they should never have been put on."

The PFAA was designed to provide protection on a block or a township basis rather than on an individual basis. The average yield within the township or block was the basis for determining whether or not there was a payout from the program. If the block average yield was below the target range then all producers within that block received a payment. It was therefore possible for a producer to receive a payout while having yields that were above the payout range. It also was possible to receive no payout with a yield which was within the payout range.

The average yield used for basing a payout was the wheat yield. Payouts were based on three different ranges:

1) An average yield of 0 to 4 bushels per acre received a $2.50 per cultivated acre payment.
2) An average yield of 4 to 8 bushels per acre received a $1.50 per cultivated acre payment.
3) An average yield of 8 to 12 bushels per acre received 10 cents for each cultivated acre for each cent (not exceeding 10) that the average price of wheat fell below 80 cents per bushel.

Payments were made on half of the farmer's cultivated land with a maximum payment of $500 in the highest payout range and a maximum payment of $400 in the second range. The minimum payment was $200 for the highest range. In order to receive the maximum payment, a producer required only 400 acres of cultivated land.

The program was funded by a one percent levy, which was matched by the federal government, on all marketed grain. Levies were not based on yields in an area nor were they used to fund the program in the area in which they were collected. Levies were the same for all regions and were not affected by yield variability or fund surplus/deficit in a particular region.

The 1940s did see prosperity begin to return to the prairies. With a rebound of wheat yields in Saskatchewan to 18 bushels per acre in 1942, there was a general improvement in farm incomes.

This trend continued into the 1950s, but began to falter shortly thereafter. By the mid 1950s, people were again beginning to question the viability of farming and living in western Canada. In 1954, wheat yields fell to 8 bushels per acre in Saskatchewan and, while not being as low as in 1937, were severe enough to renew farmers' memories of the 1930s. Wheat yields did recover somewhat after 1954, only to fall again in the period 1957 to 1959. Farmers, fearing the return of the 1930s, were not convinced that the PFAA was providing them with adequate protection.

Strong criticism of the program had developed over time and there were calls for a major reform or for the creation of a new crop insurance program. The PFAA was criticized because the levy was collected in a way that did not differentiate between regions. Therefore, areas that had higher yields with little variability were paying a levy so that farmers in other regions could receive a payout. Another criticism of the PFAA was that the use of a block or township average yield meant that individuals could suffer poor yields but receive no payout because the township or block yield was such that a payout was not triggered. Lastly, the PFAA was criticized for becoming obsolete in a new era of agriculture. The average yields and maximum acreage which were used to determine payments were no longer relevant to the agriculture that had evolved since the inception of the program. With new seed varieties and farm practices, yields were significantly higher on average than those used for payout calculations. As well, the average farm size was now considerably larger than the allowable 400 acres, making the maximum payment insignificant to most producers.

3.4 Introduction of the Canadian Crop Insurance Legislation

In 1957, the Progressive Conservatives led by the Hon. John Diefenbaker were brought to power. Their victory included a virtual sweep of the prairie seats. With a great deal of support in the west there was pressure on the government to deliver long term agriculture programs. One repeated demand from farm organizations was for a crop insurance program.

In Saskatchewan, the Royal Commission on Agriculture and Rural Life, which was completed in 1956, held 57 community hearings throughout Saskatchewan. At 16 of these hearings, the merits of a crop insurance program were discussed. Whenever crop insurance was debated, the majority of the people attending the hearings were in support of the concept. Major organizations which presented briefs to the Commission were supportive of a crop insurance program. Groups such as the Saskatchewan Wheat Pool, the Saskatchewan Board of Trade, the Saskatchewan Agricultural Societies Association, the Saskatchewan Association of Rural Municipalities, and other groups, all supported the idea of a crop insurance program.

If there was any opposition to crop insurance, it was usually based on practical and theoretical concerns. In the first place, it was argued that the premiums necessary to make the program actuarially sound would be so high that farmers would not be able to afford to join the program. Part of the attractiveness of the PFAA was the low premium. In the second place, people argued that the establishment of a crop insurance program would bring an end to the PFAA which had provided support to producers in times of low production.

Although support for crop insurance may not have been unanimous, there was sufficient support to convince the government to introduce the crop insurance legislation. On June 29, 1959, the Diefenbaker government passed the federal Crop Insurance Act. The legislation did not give the federal government the authority to operate its own crop insurance program. The country's constitutional arrangement, by which the provinces have authority over property and civil rights, prevented the federal government from running a strictly federal crop insurance program. The legislation allowed only for the payment of contributions and loans from the consolidated fund to provincial programs for crop insurance.

The original Crop Insurance Act allowed the federal government to enter into agreement with any province, in which it would provide a payment to the province for the costs incurred by the province in the operation of a crop insurance program. The Act also allowed the federal government to make loans to the province for the operation of the program. The original Crop Insurance Act committed the federal government to covering 50 percent of the administration costs of a provincial program and 20 percent of the premiums charged. Also, loans to provinces were provided for indemnity commitments of 75 percent of the required amount less $200,000.

The most controversial feature of the Act was that while the program was voluntary, any farmer signing up for crop insurance was not eligible to receive payments from the PFAA. The reasoning was that insurance coverage would only be offered to farmers under one program and not both. The federal government did not want to be in a situation of making two payments on the same crop loss.

The federal legislation also stipulated that for a province to receive payment from the federal government, the crop insurance program must be actuarially sound and have a minimum number of farmers in the program. The suggestion was that at least 25 percent of farmers or 25 percent of total acreage would have to enter into the insurance unit. The insurance unit did not have to be the whole province but could be a township or some other defined area.

The Act also stipulated that financial assistance would be provided only if the provincial program insured up to, but not more than 60 percent of the long term average yield in the area. The federal government did not want crop insurance to be a major subsidy program. By providing a minimum level of support, the federal government sought to encourage the participation of farmers in the program. For the most part, the program was designed to be financed by the farmers' contributions. In describing the program, the federal government often made comparisons between crop insurance and the federal unemployment program. The unemployment program also paid 20 percent of the premiums and it was argued that crop insurance was simply unemployment insurance for farmers.

3.5 Evolution of Crop Insurance

Since it was enacted, the federal Crop Insurance Act has undergone many changes. Most of these changes were intended to increase the benefits from crop insurance in an attempt to increase farmer participation in the crop insurance program. As such, these changes led crop insurance farther away from its stated objective of not being a major subsidy program.

In 1964, the federal Crop Insurance Act was amended to provide for a reinsurance fund. This was in response to concerns voiced by the provinces that they had to limit participation in the program because a large crop failure would require them to pay the difference between payouts and premiums collected. The magnitude of the payouts could compromise their ability to borrow money. This concern acted as a major determinant of crop insurance program development in many provinces in the 1960s.

The reinsurance fund was established so that 15 percent of total premiums went to the Reinsurance Fund of Canada and 15 percent of total premiums went to the Reinsurance Fund of the Province. Provisions were also made to allow the premiums collected for the fund to vary according to the financial status of the fund. Payouts from the reinsurance fund were made when insurance claims exceeded premiums collected. The provincial reinsurance fund would pay, on the first amount owed, a sum equivalent to 2.5 percent of total indemnities. Any remaining amount was then split with 75 percent being covered from the federal reinsurance fund and 25 percent from the provincial reinsurance fund. This greatly reduced the province's exposure in the event of widespread crop failure.

In 1966, the federal Crop Insurance Act was again amended in an attempt to increase farmer participation in the program. The coverage level was increased from 60 percent of the long term average yield to 80 percent of the long term average yield. The second major change was to increase the federal contribution of premiums to 25 percent from 20 percent. In 1970, minor amendments were made to the Act to allow for expanded coverage to cover all losses resulting from a farmer's inability to seed a crop because of weather conditions.

In 1973, the federal government amended the Crop Insurance Act to allow for alternative funding arrangements of provincial crop insurance schemes. Under one funding arrangement, both the federal and provincial governments would each contribute 25 percent of the total premium costs and 50 percent of the total provincial administration costs. Under the second funding arrangement, the federal government would pay 50 percent of the total premium cost and the province would pay 100 percent of the administration cost. Either funding arrangement meant that the producer would only have to pay 50 percent of the total crop insurance premium. The other significant change made at this time was the inclusion of hail spotloss-coverage. The result of this was that for many producers the cost of buying crop insurance with hail protection was comparable to purchasing just hail insurance from a private hail insurance company.

3.6 Crop Insurance Program—Saskatchewan

Each province operates its own crop insurance program within the framework established by the federal legislation. As such, the operations and guidelines of each crop insurance program vary between provinces. However, because the major features of crop insurance programs are similar in all provinces, a brief description of the Saskatchewan crop insurance program will serve as a general overview of all provincial crop insurance programs.

Saskatchewan's crop insurance program is a voluntary all-risk crop insurance program. Coverage is offered for crop losses due to natural hazards such as drought, flood, frost, snow, and grasshoppers. The program guarantees that producers of eligible crops will receive a payment equal to the revenue from a minimum yield at a predetermined price. The Saskatchewan program provides coverage on over 20 types of crops. The following description is for general insurance coverage only and does not include coverage for hail, reseeding, and other program features.

A producer must purchase insurance on a crop basis. In order to insure a crop, coverage must be purchased on all the acreage seeded to that crop. Payouts are calculated on the farm's average yield for that crop rather than on the average yield for a specific parcel of land. Adjustments are also made for quality loss below base grade for each crop. The province is divided into 23 different risk areas for the purpose of calculating coverage and premiums.

Producers have the option of purchasing two different types of yield coverage: area-based, or individual coverage. Area-based coverage offers producers yield insurance protection on eligible crops at 60 to 70 percent of the 15 year average yield in their risk area. This coverage is then adjusted by a productivity index to account for the land quality difference of each quarter section. Individual coverage allows producers to guarantee their production at a base coverage of 80 percent of the actual 10 year average for their own farm. This describes the coverage which producers receive in the first year of purchasing crop insurance. With area-based coverage, in subsequent years, the guaranteed yield is adjusted to between 70 and 125 percent of the base coverage, depending on the accumulated loss ratio for that crop. This rewards producers who make no claims against the program by increasing their production guarantee, and penalizes producers who make successive claims against the program. The production guarantee can then be expressed as:

$$\text{Production Guarantee} = (\text{Base Coverage} \times \text{Coverage Adjustment}). \tag{1}$$

Producers have the choice between guaranteeing their yields at a high or low price, which is based on the expected market returns. Another available price option is based on projected market returns for that crop. The total cost of the insurance is a percentage of the value of the production guarantee and can be expressed as follows:

Total Cost = (Production Guarantee) × Unit price × Premium Rate. (2)

The percentage used for the premium rate is based on the last 25 years of yield variation in that area and measures the probability of loss. Producers receive an indemnity payment when their actual yield is less than the production guarantee. The indemnity or crop payments are calculated as:

Indemnity = (Production Guarantee - Total Yield) × Unit Price. (3)

3.7 Overview of Crop Insurance Performance

Although the federal government brought in crop insurance legislation in 1959, the provinces were slow to implement the program. Manitoba, in 1960, was the first to introduce a program, followed shortly by Saskatchewan. By 1965, only 828,000 acres were insured. Most of the insured acres at this time were in the three western provinces. The initial low participation was a reflection of the low subsidy element in the program and the relative newness of crop insurance. Even in the prairie provinces, crop insurance was not available to all producers. The stipulation in the federal legislation, that 25 percent of the producers or 25 percent of the land area must be registered for crop insurance before it could be implemented, restricted the areas in which the province could introduce the program. The level of protection offered by crop insurance also deterred participation since the coverage was only 60 percent of the long term yield.

Over time, participation in crop insurance grew. By 1968, over 9 million acres had been enrolled in crop insurance programs (Tables 1 and 2). The growth in participation was in part the result of all the provinces introducing crop insurance programs. The majority of the acres enrolled in the program, however, were

Table 1. Historical review of Canada crop insurance (thousands)

Year	Acres	Liability $	Total Premium $	Indemnities $	Cumulative balance in fund $
61-62	31	313	32	129	-97
62-63	226	2296	219	75	46
63-64	439	3547	354	23	376
64-65	465	4128	366	152	590
STOTAL	1163	10284	972	381	590
65-66	827	8034	685	217	467
66-67	2341	25792	1996	453	2010
67-68	6276	88247	6680	2732	5958
68-69	9346	136822	10726	10497	6187
69-70	8091	126894	9873	14999	1062
70-71	5334	106182	11085	5824	6323
71-72	6338	119037	9116	19411	-3972
72-73	7318	190339	10602	9837	-3206
73-74	14589	361556	31487	18393	9887
STOTAL	60463	1162908	92254	82367	9887
74-75	18211	640246	59918	62848	-2930
75-76	22710	941598	91064	61624	26509
76-77	23782	1121227	109554	60742	75322
77-78	29262	1457792	147510	107677	115154
78-79	27759	1488957	141616	73739	183031
79-80	27206	1588661	145306	201653	126683
80-81	30647	2115454	191920	259787	58816
81-82	33942	2586375	238582	166674	130725
82-83	33272	2795457	286862	257913	159673
83-84	34523	3043595	286955	284890	161738
84-85	36328	3252650	321352	538702	-55612
85-86	40987	3790231	390620	711376	-376367
86-87	44305	4128314	440653	390635	-326349
87-88	41090	3115661	344029	263984	-246304
88-89	42715	3410812	386028	831296	-691571
STOTAL	486748	35477034	3581975	4273547	-691571
Total	548375	36650227	3675202	4356297	-681094

Source: Canadian Crop Insurance Annual Report, various issues: 1965-1988, Ottawa.

Table 2. Historical review of crop insurance—provinces (thousands)

Province	Year	1970-71	1975-76	1980-81	1985-86	1988-89	Subtotal 1965-74	Subtotal 1975-90	Total
Alberta	Acres	2510	5112	6777	9540	11666	25289	112505	137794
	Liability	35111	215955	388979	774392	791510	521212	7407877	7929089
	Total premium	3181	27366	54229	98932	139378	43804	1065322	1109126
	Indemnity	2019	17869	33210	217331	156420	49069	1202528	1251598
British Columbia	Acres	73	185	143	283	142	755	2680	3436
	Liability	7500	22080	40683	80998	87888	60220	767241	827462
	Total premium	560	2182	3463	8587	9540	4828	76899	81728
	Indemnity	219	400	2393	13839	8214	8623	86696	95320
Saskatchewan	Acres	1004	12740	15986	20754	21222	31208	245771	276979
	Liability	6868	405090	93584	1567793	1165828	553151	14691496	15244647
	Total premium	799	41353	82955	153577	126522	54718	1419805	1474523
	Indemnity	104	25742	135389	364672	465013	42768	1919726	1962494
Manitoba	Acres	1565	3635	4741	5621	5592	18282	68199	86482
	Liability	26673	126389	245847	452445	394770	343141	4071007	4414149
	Total premium	4899	9269	19932	44800	40544	28892	368619	397512
	Indemnity	1964	9529	54933	38604	120459	21182	414783	435966
Ontario	Acres	127	647	1110	1375	1908	1663	16851	18515
	Liability	7649	132422	309176	373169	415878	187875	4170599	4358475
	Total premium	551	7928	20931	34189	33572	11491	352116	363607
	Indemnity	259	6451	14619	10535	42379	10894	322386	333281
Quebec	Acres	32	324	1803	1902	2098	1220	21271	22492
	Liability	20956	26015	161955	375560	500452	106345	3183444	3289790
	Total premium	1022	2021	7724	23885	31251	6636	191761	198398
	Indemnity	1082	622	16475	7483	36696	9121	212334	221455
Maritimes	Acres	25	67	87	104	87	255	1255	1510
	Liability	1425	13646	32976	58764	54486	31277	545122	576399
	Total premium	72	945	2684	5605	5220	1801	47532	49333
	Indemnity	176	1011	2769	6846	2116	3554	56200	59754

located in the three prairie provinces. Alberta led the way with 2.5 million acres in 1970-71, followed by Manitoba with 1.6 million acres, and Saskatchewan with 1.0 million acres. The increased participation in crop insurance was also a result of the federal government's amending its legislation to increase coverage to 80 percent of the long term yield, from 60 percent, and increasing its share of the premium to 25 percent, from 20 percent.

Participation dropped off in the early 1970s, falling as low as 5.3 million acres in 1971. This drop in participation reflects the large grain surpluses that existed at that time and the impact of the federal Lower Inventories For Tomorrow program which removed land from production. However, with the federal government increasing its share of the premiums to 50 percent and including hail spotloss-coverage in 1973, participation increased dramatically. In 1974, participation was 14.6 million acres, and by 1981 it was 30.6 million acres. The province of Saskatchewan was the largest participant in the program with 16 million acres enrolled in 1981, compared to Alberta with 6.7 million acres and Manitoba with 4.7 million acres. In the 1980s, the eastern provinces also increased their use of crop insurance with Quebec having 1.8 million acres and Ontario 1.1 million acres enrolled in the program.

In total, crop insurance programs have paid out over $4 billion in indemnity claims with the majority of these claims being made in the last 15 years. With the prairie provinces having the vast majority of the acres enrolled in the program, they have also received the majority of the payments. Saskatchewan has received $2 billion or half of the payments while Alberta has received $1.2 billion and Manitoba $0.44 billion.

3.8 Financial Analysis

The previous section showed that the majority of crop insurance benefits flowed to the prairie provinces. Aggregate calculations, however, inflate the benefits received by those provinces which had high enrollments in the crop insurance programs. It is the per acre benefits that need to be examined when comparing provincial programs. In this section, loss ratios will be used first to describe the distributional impact between provinces, and secondly to indicate the level of subsidies which exist in the insurance program. The loss ratio is the ratio of indemnities to premiums. A ratio of one indicates that for every dollar of payouts or indemnities there was an offsetting dollar of premiums collected. For Canada as a whole, the loss ratio from 1959 to 1988 is 1.19 (Fig. 1). That is, for every dollar of premium collected, $1.19 were paid out in indemnities. The province of Ontario is the only province with a loss ratio less than one. Saskatchewan has the highest loss ratio at 1.33 followed by British Columbia at 1.17 and Alberta at 1.13.

The loss ratio as calculated above, is not the same as the loss ratio used by private crop insurance companies. This loss ratio does not account for the subsidy component in the premiums and there are no administration costs charged to the program. If crop insurance was offered in the private market, these factors would be taken into account. To calculate the true subsidy which exists in the program and the distribution of this subsidy between the provinces, the private loss ratio is used. For Canada, the private loss ratio for 1959 to 1988 is 2.83 (see Fig. 1). This means that for every dollar of premiums, there are $2.83 of benefits received, or a subsidy element of 183 percent.

The private loss ratio is also used to determine which provincial crop insurance program has the largest subsidy component. Quebec has the highest level of subsidy with a loss ratio of 3.25 or a subsidy level of 225 percent. The province with the next highest subsidy component is Alberta followed by Saskatchewan, British Columbia, the Maritimes, Manitoba, and Ontario.

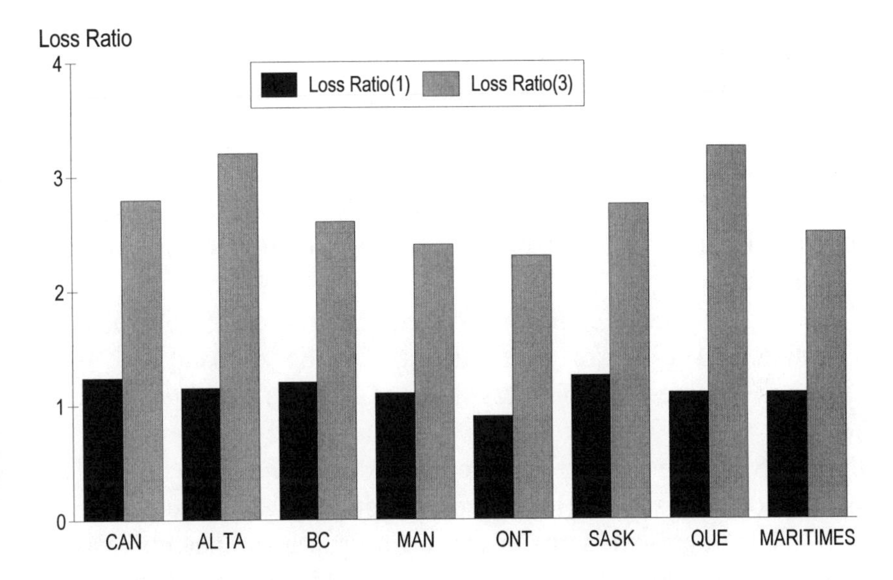

Fig. 1. Canadian crop insurance, cumulative loss ratio 1959-1988 by province

Note: Loss Ratio(1) = Indemnity/Total Premium
Loss Ratio(3) = (Indemnity + Administration Cost)/Farm Premium

3.8.1 Competitive Crop Insurance Market. Crop insurance is a conditional contract under which a farmer pays the cost of the protection (premium), with his crop output determining the indemnity. Crop insurance provides a production guarantee to the producer. When due to natural perils, the actual output for a crop falls below the production guarantee, producers receive indemnity payments. In a competitive crop insurance industry, market equilibrium requires that the indemnities and premiums are in balance in the long term. That means that the loss ratio (indemnities plus all administration costs divided by premiums) should equal one.

Fig. 2 illustrates the market equilibrium condition for crop insurance. The major determinants in the demand for crop insurance (D) are the producers' utility for income and the price of the insurance. The intersection of the aggregate demand and supply (S) curves for crop insurance determine the equilibrium quantity of insurance (Q) and the corresponding premium (P).

The aggregate demand curve for crop insurance consists of a range of producers with differing risk preferences. As the price for crop insurance declines, more risk-neutral or risk-taking producers purchase insurance. At the higher price levels, only risk-averse producers purchase crop insurance. If more producers are to participate in crop insurance, the cost of crop insurance must be reduced. This is shown by the shifting of the supply curve from S to S′.

The shifting of the supply curve can be achieved by the government operating the program and not charging administration to the program, or by not operating the program at an actuarially sound level (charging premiums less than that indicated by the private loss ratio). The result of this is that the premium the producer must pay (P1) is less than the market insurance rate of P. Participation then increases from Q to Q1. The slope of the demand curve (D) will determine the increase in participation due to the subsidy of P2-P1.

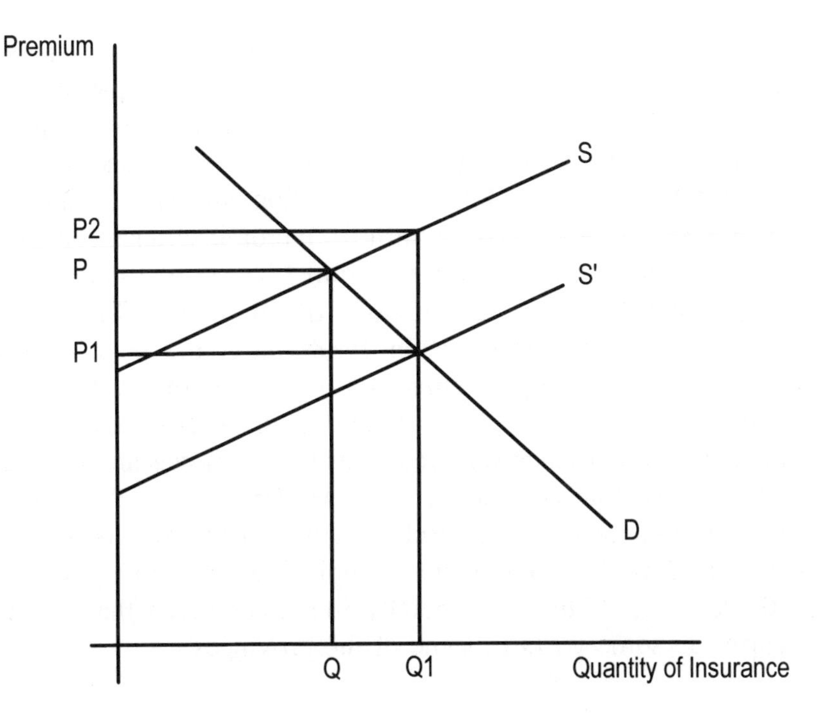

Fig. 2. Crop Insurance market with government involvement

3.8.2 Impact of Subsidy Level and Participation. During the 31 years that crop insurance programs have operated in Canada, participation has grown rapidly. The percentage of producers participating in the program has risen from 26% in 1966 to 65% in 1986. As mentioned in the previous section, the federal and provincial governments have provided a subsidy of around 183 percent to the crop insurance program. With this high level of subsidy, farmers, on average over time, receive more from the program than they contribute. However, there is still the empirical question as to why many producers choose not to join.

3.8.3 Empirical Estimation of the Demand for Canadian Crop Insurance. To understand the variation in participation rates, data were obtained on crop insurance participation by crop and province over the history of the program. The demand for crop insurance, as represented by the participation rate, can be expressed as a function of the expected return from crop insurance (Table 3).

Our data covers five years, from 1984-85 to 1988-89, with all aggregate data over the previous fifteen years. Table 3, for different provinces and crops in 1988 shows the percentage of acreage insured, farm premiums as a percentage of liabilities, and return as a percent of liability. Since individual farm data were not available, each province was treated as though it was an individual farm. Our sample consists of 30 observations from 9 provinces covering the major crops in each province. Data was used over a five year period to avoid any sample bias due to selecting a particular year. This is especially important given that drought occurred in some years in some of the provinces.

3.8.4 Percentage of Acreage Insured. The dependent variable, the quantity of insurance, is the percentage of acreage insured. In our sample, it varies from 1.19% for grain corn in Ontario in 1987, to 94.4% for grain corn in Manitoba in 1987.[4]

3.8.5 Total Farm Premiums As Percentage of Liability. The total farm premium as a percentage of liability is calculated by crop for each province. In 1973, the federal government agreed to pay 50% of the total premium cost with the provinces paying 100 percent of administration costs. Therefore, the cost of insurance for the farmer is 50% of the total premium. The farm premium as a percentage of liability is the total farm premium divided by the liability. Its value varies from a high for oats of 18.67% in British Columbia in 1987, to a low for grain corn of 2.2% in Ontario in

[4]For information on years other than 1988-89 see appendix A.

Table 3. Empirical estimation of the demand for Canadian crop insurance, 1988-89

Province	Crop	Percentage of acreage insured	Total farm premium as % of liability	Expected return as % of liability	Expected rate of return	Coefficient variance on yield
Alberta	Barley	61.78	7.86	14.63	0.86	0.22
	Canola	74.26	11.60	19.91	0.72	0.13
	Oats	27.74	9.91	17.12	0.73	0.10
	Wheat	67.92	7.61	13.89	0.83	0.13
British Columbia	Barley	20.24	9.24	12.10	0.31	0.23
	Canola	33.79	8.39	24.23	1.89	0.15
	Oats	6.00	15.67	18.27	0.17	0.14
	Wheat	25.91	9.10	16.24	0.78	0.16
Manitoba	Barley	50.54	3.86	5.75	0.49	0.26
	Flax	57.34	5.62	8.77	0.56	0.24
	Oats	45.15	5.98	10.95	0.83	0.16
	Wheat	54.81	4.13	5.99	0.45	0.17
	Grain corn	64.05	9.05	24.70	1.73	0.23
Maritime	Oats	5.53	6.20	8.55	0.38	0.13
	Mixed grain	3.88	4.81	9.56	0.99	0.17
	Wheat	16.50	4.21	10.33	1.45	0.15
	Barley	9.74	5.41	8.55	0.58	0.25
Ontario	Grain corn	1.28	2.20	4.17	0.90	0.09
	Soybeans	56.10	2.71	7.16	1.64	0.13
	Wheat	45.89	2.75	3.59	0.31	0.15
Quebec	Barley	48.15	2.58	5.05	0.96	0.27
	Oats	10.18	3.76	4.56	0.21	0.18
	Wheat	31.96	3.75	8.12	1.17	0.21
	Grain corn	62.95	6.66	8.59	0.29	0.11
Saskatchewan	Barley	90.75	5.42	12.25	1.26	0.23
	Canola	58.91	7.04	12.97	0.84	0.15
	Oats	77.43	7.86	17.92	1.28	0.13
	Wheat	75.32	4.80	9.93	1.07	0.14
	Mustard	74.41	8.77	15.97	0.82	0.16
	Flax	85.03	6.24	13.80	1.21	0.20

1988.

3.8.6 Expected Return as Percent of Liability. The expected rate of return is defined as the average ratio of indemnities to liability. For 1984, the expected return is calculated from 1973 to 1984. For 1985 to 1988, one previous year's data was added into the calculation. In other words, for 1984, a 10-year average ratio is used, and for 1985 an 11-year average ratio is used, and so on. The previous year's data is added into our calculation, since farmers will calculate their expected rate of return for insurance with the most up-to-date information. For example, in 1988, they obtain the expected return by calculating the average ratio from 1973 to 1987, instead of from 1973 to 1986.

3.8.7 Expected Rate of Return. The expected rate of return is one of the independent variables. It is calculated as the average ratio of indemnities to premiums, minus one. With a higher expected return to premiums ratio, we expect to have a higher participation rate. A positive value of expected rate of return indicates that farmers get more from the crop insurance program than they pay. In that sense, farmers should be more willing to participate in the program. Therefore, the demand for crop insurance and the expected rate of return should be positively related. The average level of subsidy in Canadian crop insurance is indicated by the private loss ratio of 1.83. However, this varies between provinces and crops. Expected rates of return range from -0.03 for barley in British Columbia in 1984 to 2.81 for grain corn in Quebec in 1987. The expected rate of return indicates that producers of barley in British Columbia could expect to get back only 97 cents for every dollar of crop insurance they purchase. Grain corn producers in Quebec could expect to get back $3.81 for every dollar of crop insurance they purchase.

The results of the regression analysis of participation rates is shown in Table 4. The coefficient of the expected rate of return

Table 4. Regression coefficients explaining Canada crop insurance participation, 1984-1988

	Regression equation	
Independent variables**	(1)	(2)
Expected rate of return	18.54 (5.14)	18.24 (5.08)
Yield coefficient of variation	61.90 (1.68)	
Constant	26.40 (7.43)	16.00 (2.24)
R2	0.15	0.17

**t statistics in parentheses
Dependent variable is percentage of acreage insured.

indicates that the participation rate is influenced by the level of return or subsidies. The coefficient of 18.54 in the regression equation (1) implies that an increase in the expected rate of return of 10 percent would increase participation by 1.85 percentage points. Our results are very similar to those obtained by Gardner and Kramer (1986) in their study of U.S. crop insurance.

While the expected rate of return is significant in determining the participation rate, the low coefficient of variation would indicate that other factors must also affect participation. In equation (2), another independent variable was introduced, the yield coefficient of variation. The yield coefficient of variance measures the variability in yields over a standardized mean.

A greater yield coefficient of variation is expected to increase the demand for the program. A high variation in yields implies greater uncertainty with respect to farm income; therefore, farmers would be more willing to buy crop insurance to stabilize their income. Thus, it should have a positive relation to the participation rate.

The results of regression equation (2) indicate that participation rates are positively related to expected rate of return to insurance and to the yield coefficient of variation. The coefficients are 18.24 and 61.6 respectively. This implies that the variability in yields has a greater impact on participation than the level of subsidy of the program itself.

The correlation coefficient is still relatively low which could indicate some difficulty in using the yield coefficient of variance as a measure of the riskiness in producing a crop. The coefficient does not enable one to distinguish between the variance around a flat mean yield and a yield on an upward trend. In the latter case, technology could be outpacing crop insurance yield coverage, making crop insurance inadequate for the farmer.

The low correlation coefficient may also be partly explained by the use of aggregate data as well as by differences between the different provincial crop insurance programs. As well, crop insurance demand is dependent on other factors such as age, education, farm size and debt.

3.9 Summary

The introduction of crop insurance in Canada in 1959 was the result of farmer pressure for a long term insurance program. The government, in implementing the program, did not intend it to be a major subsidy program. Rather, they saw it as being a similar program to unemployment insurance which was available to Canadian workers. However, a large subsidy factor is present. In total, Canadian crop insurance is subsidized by 183 percent by both the federal and provincial governments.

The development of crop insurance in Canada has strayed considerably from its goal of the 1960s of receiving only a minor subsidy. After the first 5 years of operation, crop insurance was truly a program without a subsidy component. The private loss ratio indicates that the program was making a 9 percent profit with that level of government participation. Crop insurance participation, however, was deemed to be too low and governments began to increase their share of contributions to the program.

As the result of the increased involvement of governments in crop insurance, farmer participation increased. This is consistent with our analysis on the effect of subsidies and participation rates. The results indicate that producers will participate in crop insurance if sufficient subsidies are offered. Our results also suggest, though, that other factors must be important in influencing producers' decisions on whether or not they join the program. The low correlation coefficient in the results seems to suggest this. Perhaps this is why an average subsidy of 183 percent has not attracted higher participation levels.

Interestingly, the province of Saskatchewan, which has the highest participation and indemnity payments, does not have the highest subsidy component to their program. The provinces of Quebec and Alberta both have subsidy levels higher than Saskatchewan. Their subsidy levels are over 200 percent.

Appendix

Table A.1. Empirical estimation of the demand for Canadian crop insurance, 1984-1985

Province	Crop	Percentage of acreage insured	Total farm premium as % of liability	Expected return as % of liability	Expected rate of return	Coefficient variance
Alberta	Barley	52.40	6.39	9.00	0.41	0.22
	Canola	61.49	8.39	13.68	0.63	0.13
	Oats	27.92	8.88	13.21	0.49	0.10
	Wheat	58.14	5.41	7.00	0.29	0.13
British Columbia	Barley	15.19	7.34	7.00	-0.05	0.23
	Canola	22.06	8.94	13.67	0.53	0.15
	Oats	12.24	9.01	12.85	0.43	0.14
	Wheat	21.48	6.83	6.12	-0.10	0.16
Manitoba	Barley	49.69	3.62	6.68	0.85	0.26
	Flax	49.96	5.85	10.37	0.77	0.24
	Oats	25.99	5.32	10.62	1.00	0.16
	Wheat	51.03	3.42	5.57	0.63	0.17
	Grain corn	54.09	6.33	19.45	2.07	0.23
Maritime	Oats	8.46	4.98	8.73	0.75	0.13
	Mixed grain	6.67	5.15	7.44	0.45	0.17
	Wheat	12.95	5.06	12.33	1.44	0.15
	Barley	15.42	5.07	6.90	0.36	0.25
Ontario	Grain corn	1.62	3.01	4.47	0.49	0.09
	Soybeans	43.52	3.57	5.49	0.54	0.13
	Wheat	31.85	4.12	5.93	0.44	0.15
Quebec	Barley	27.58	3.08	6.87	1.23	0.27
	Oats	3.47	4.12	7.71	0.87	0.18
	Wheat	13.21	4.18	7.90	0.89	0.21
	Grain corn	27.67	3.89	6.37	0.64	0.11
Saskatchewan	Barley	57.14	4.67	10.02	1.15	0.23
	Canola	55.52	6.31	14.26	1.26	0.15
	Oats	44.34	6.49	14.60	1.25	0.13
	Wheat	67.87	4.25	8.29	0.95	0.14
	Mustard	65.68	7.69	17.27	1.25	0.16
	Flax	57.52	6.43	13.88	1.16	0.20

Table A.2 Empirical estimation of the demand for Canadian crop insurance, 1985-1986

Province	Crop	Percentage of acreage insured	Total farm premium as % of liability	Expected return as % of liability	Expected rate of return	Coefficient variance
Alberta	Barley	63.58	6.87	12.95	0.89	0.22
	Canola	62.8	8.64	16.7	70.94	0.13
	Oats	33.41	9.26	15.54	0.68	0.10
	Wheat	65.62	5.93	10.38	0.75	0.13
British Columbia	Barley	41.09	8.79	8.57	-0.03	0.23
	Canola	40.30	0.70	16.75	0.57	0.15
	Oats	33.23	11.42	15.00	0.31	0.14
	Wheat	40.67	7.23	7.86	0.09	0.16
Manitoba	Barley	53.93	4.18	6.69	0.60	0.26
	Flax	54.21	6.41	10.21	0.59	0.24
	Oats	29.58	6.23	11.26	0.81	0.16
	Wheat	56.12	4.57	5.48	0.20	0.17
	Grain corn	63.75	6.90	18.58	1.69	0.23
Maritime	Oats	7.21	5.38	10.16	0.89	0.13
	Mixed grain	5.94	5.36	9.9	0.85	0.17
	Wheat	18.05	5.20	11.22	1.16	0.15
	Barley	14.73	5.45	9.37	0.72	0.25
Ontario	Grain corn	1.75	3.16	4.31	0.36	0.09
	Soybeans	43.40	3.45	4.86	0.41	0.13
	Wheat	31.31	3.73	5.58	0.50	0.15
Quebec	Barley	36.10	2.57	5.99	1.34	0.27
	Oats	3.22	4.09	7.44	0.82	0.18
	Wheat	19.08	3.38	6.80	1.01	0.21
	Grain corn	29.25	3.28	5.15	0.57	0.11
Saskatchewan	Barley	62.60	4.99	11.99	1.41	0.23
	Canola	61.40	6.43	15.04	1.34	0.15
	Oats	52.20	7.09	17.27	1.44	0.13
	Wheat	72.92	4.54	9.78	1.16	0.14
	Mustard	89.03	7.91	22.68	1.87	0.16
	Flax	61.57	6.03	15.96	1.65	0.20

Table A.3. Empirical estimation of the demand for Canadian crop insurance, 1986-1987

Province	Crop	Percentage of acreage insured	Total farm premium as % of liability	Expected return as % of liability	Expected rate of return	Coefficient variance
Alberta	Barley	58.83	8.30	16.01	0.93	0.22
	Canola	68.89	9.69	18.75	0.93	0.13
	Oats	37.05	9.85	17.94	0.82	0.10
	Wheat	69.32	7.24	13.43	0.86	0.13
British Columbia	Barley	61.30	9.90	11.49	0.16	0.23
	Canola	73.72	12.43	23.03	0.85	0.15
	Oats	15.10	13.61	18.55	0.36	0.14
	Wheat	62.84	8.20	16.01	0.95	0.16
Manitoba	Barley	52.83	4.11	6.33	0.54	0.26
	Flax	58.19	6.00	9.51	0.59	0.24
	Oats	30.38	6.52	11.26	0.73	0.16
	Wheat	56.81	4.37	5.08	0.16	0.17
	Grain corn	85.05	7.24	26.76	2.70	0.23
Maritime	Oats	8.08	5.63	9.79	0.74	0.13
	Mixed grain	5.72	5.64	9.74	0.73	0.17
	Wheat	18.80	5.08	11.10	1.19	0.15
	Barley	14.99	5.64	9.47	0.68	0.25
Ontario	Grain corn	1.48	2.78	4.14	0.49	0.09
	Soybeans	46.28	3.10	4.48	0.45	0.13
	Wheat	29.49	3.50	4.89	0.40	0.15
Quebec	Barley	40.88	2.31	4.39	0.90	0.27
	Oats	4.33	3.46	6.50	0.88	0.18
	Wheat	28.07	3.04	4.72	0.56	0.21
	Grain corn	44.58	2.72	5.65	1.08	0.11
Saskatchewan	Barley	67.50	5.22	12.92	1.48	0.23
	Canola	66.14	6.68	13.57	1.03	0.15
	Oats	58.51	7.61	19.20	1.52	0.13
	Wheat	79.65	4.52	12.10	1.68	0.14
	Mustard	77.37	7.73	25.55	2.31	0.16
	Flax	66.85	6.15	16.10	1.62	0.20

Table A.4. Empirical estimation of the demand for Canadian crop insurance, 1987-1988

Province	Crop	Percentage of acreage insured	Total farm premium as % of liability	Expected return as % of liability	Expected rate of return	Coefficient variance
Alberta	Barley	56.81	7.84	14.94	0.91	0.22
	Canola	65.28	11.11	19.23	0.73	0.13
	Oats	33.52	10.02	17.30	0.73	0.10
	Wheat	65.39	7.66	13.28	0.73	0.13
British Columbia	Barley	48.05	10.58	11.79	0.11	0.23
	Canola	62.39	14.26	23.62	0.66	0.15
	Oats	14.90	18.67	18.80	0.01	0.14
	Wheat	55.81	10.16	17.80	0.75	0.16
Manitoba	Barley	50.98	4.01	6.22	0.55	0.26
	Flax	53.49	5.77	9.02	0.56	0.24
	Oats	31.74	6.28	11.05	0.76	0.16
	Wheat	53.76	4.30	5.36	0.25	0.17
	Grain corn	94.40	9.24	26.07	1.82	0.23
Maritime	Oats	2.10	5.58	9.62	0.72	0.13
	Mixed grain	5.52	5.51	9.58	0.74	0.17
	Wheat	14.38	5.05	11.06	1.19	0.15
	Barley	13.10	5.72	9.17	0.60	0.25
Ontario	Grain corn	1.19	2.51	4.27	0.70	0.09
	Soybeans	50.55	2.91	4.10	0.41	0.13
	Wheat	30.67	3.20	4.98	0.56	0.15
Quebec	Barley	43.86	2.35	3.79	0.61	0.27
	Oats	6.09	3.54	6.98	0.97	0.18
	Wheat	32.18	3.43	4.29	0.25	0.21
	Grain corn	54.30	3.34	12.71	2.81	0.11
Saskatchewan	Barley	64.62	5.42	12.27	1.27	0.23
	Canola	61.72	6.80	14.13	1.08	0.15
	Oats	56.47	7.99	18.42	1.31	0.13
	Wheat	74.89	4.93	11.23	1.28	0.14
	Mustard	54.98	13.39	16.45	0.23	0.16
	Flax	57.61	6.38	15.98	1.50	0.20

References

Gardner BL, Kramer RA (1986) Experience with crop insurance programs in the United States. In: Hazell P, Pomareda C, Valdes A (eds) Crop insurance for agricultural development: issues and experience. Johns Hopkins Univ Press, Baltimore, London

Halcrow HG (1949) Actuarial structures for crop insurance. J Farm Econ 31:418-43

Hansard. House of Commons: August 30, 1958, pp. 4345-4353

Hansard. House of Commons: June 29,1958, pp. 5222-5255

Hansard. House of Commons: July 7, 1959, pp. 5582-5637

Hansard. House of Commons: October 6, 1964, pp. 8808-8824

Hansard. House of Commons: October 13, 1964, pp. 8976-8990

Hansard. House of Commons: June 20, 1966, pp. 6667-6671

Hansard. House of Commons: June 21, 1966, pp. 6701-6707

Hansard. House of Commons: July 7, 1966, pp. 7335-7356

Hansard. House of Commons: September 16, 1964, pp. 8101-8108

Standing Committee on Agriculture and Colonization. Minutes of Proceedings and Evidence. House of Commons: June 10, 1947, Ottawa

Royal Commission on Agriculture and Rural Life: Volume II, 1956

Source of Data

Canadian Crop Insurance Annual Report, various issues: 1965-1988, Ottawa

Canada-Saskatchewan Crop Insurance Annual Report, various issues: 1976-1988

Statistic Canada: Field Crop Reporting Series 22-002

Canadian Grains Industry: Statistical Handbook, various issues. Canada Grains Council: Manitoba

Chapter 4

All-Risk Crop Insurance: Lessons From Theory and Experience

B.D. WRIGHT[1] and J.A. HEWITT[2]

> A producer is generally willing to use more resources, including borrowing more capital for production, if risk can be reduced. One of the major reasons for government involvement in agriculture is to ensure an abundant supply of food and fiber products at reasonable prices. To achieve that goal, the government must assist farmers in managing their risks. The Federal crop insurance program and the various other agricultural programs were designed to assist farmers in the management of risks (U.S. Congress, 1989, p. 57).

Relative to most other producers of goods and services, many agricultural producers are subject to large short-run fluctuations in their outputs and the prices they receive. These fluctuations can be major threats to financial survival, and managing them is a serious challenge for agricultural producers.

These facts are easily understood by non-farmers. Perhaps that helps explain the universal tendency to support various farm programs, like public all-risk crop insurance, that claim to give essential risk management assistance to producers, as in the passage quoted above. Serious attempts to provide such crop insurance have been made, with major financial commitments, in many developed and developing countries over a long period of time.

---

[1]Department of Agricultural and Resource Economics, University of California, Berkeley, California 94720, USA.

[2]Department of Agricultural and Resource Economics, Montana State University, Bozeman, Montana, USA.

In this paper we assess the general economic argument for all-risk or multi-peril crop insurance. In doing this we find ourselves in a very difficult position. One would hope that implimentation of such a costly government program would routinely be preceded by a formulation of objectives and a careful ex ante evaluation. But there is very little in the way of formal economic argument or ex ante evaluation available to justify the design, implementation or continuation of the, often extremely costly, all-risk crop insurance programs that exist in many countries today. Even the elementary requirement of a clear statement of objectives is typically unavailable. (The best we could come up with is the introductory quote above; see U.S. General Accounting Office 1989 for eight criteria which disaster assistance programs should meet. Crop insurance programs in the United States meet three of these criteria.)

The lack of serious prior economic evaluation would not be bothersome if farmers, taxpayers and policymakers were uniformly happy with the many public all-risk programs that have been implemented. But the worldwide experience with these programs has, in fact, been almost uniformly negative. Even though the objectives of the programs are unclear, their overall failure as reasonable public policy is all too obvious.

In this paper we provide an overview of the experience of a number of crop insurance programs in many countries. We do not consider the Canadian case, which is examined in detail in the following chapter. Against this background of historical experience, we assess the current theoretical and practical arguments for expanded crop insurance in the United States.

In our arguments, we do not consider the direct evidence regarding the nature and extent of farmers' risk aversion. These issues, despite a great deal of research, are far from resolved, and questions raised by new departures from traditional expected utility theory will take time for the profession to answer. Rather, we assume some degree of farmer risk aversion with respect to consumption fluctuations. However, risk aversion is necessary, but far from sufficient, to justify a crop insurance program.

We argue that there is too little connection between the theoretical arguments used to support public all-risk crop insurance programs, the actual economic environment of the farmer in the absence of such a program, and the experience with implementation of such programs. Here we try to bring these three aspects of the issue together. Our focus is on the policy question: Do theory, at its current state of development, and the empirical evidence, support the implementation, continuation or expansion of all-risk crop insurance?

We begin with a review of the history of all-risk insurance programs. Then we critically discuss the standard theoretical approach to crop insurance in section 4.2. The crucial distinction between crop insurance and income insurance is emphasized in section 4.3, and private risk management strategies, neglected in standard discussions, are noted in section 4.4.

The issue of who ultimately gains from a crop insurance program is relatively neglected in most discussions. In section 4.5 we argue that crop insurance will be unlikely to change farmers' welfare very significantly in the long run, due to capitalization of any benefits in asset prices. The risk-management strategies reviewed in section 4.4 are discouraged by public crop insurance. We argue in section 4.6 that, if the private strategies are more efficient, as is very likely, substitution of public insurance induces a previously unrecognized cost of moral hazard.

In the United States, the major current justification used by policymakers for expanded crop insurance is that it could eliminate the need for more costly disaster payments. The evidence on this from United States' experience and farmers' attitudes is not encouraging, as shown in section 4.7. Conclusions are summarized in section 4.8.

4.1 Historical Overview

In this section we briefly review the history of all-risk insurance and related programs. The yardstick by which success of an insurance scheme is routinely measured, is known as the loss ratio. Loss ratios are generally calculated as indemnities divided by premiums. Private insurance underwriters typically design their programs to achieve an average loss ratio of no more than 0.7 (Patrick, Lloyd and Cary 1985; Lloyd and Mauldon 1986). This is a long run goal with the purpose of achieving a reasonable return to the insurer after administrative costs are paid, and investment income is taken into account.

Note, however, that in the average year, the loss ratio must be much less than 0.7, because the distribution of indemnities is highly skewed; in disaster years indemnities greatly exceed premiums. In the good years, premiums should be collected so as to add to the capital fund, such that a balance large enough to sustain a major disaster is accumulated.

Every multi-peril insurance program that has progressed beyond infancy has been underwritten by a government. To our knowledge, attempts by private underwriters to provide multi-peril insurance have all failed (for the U.S. experience see the historical survey of Kramer (1983), for Australia, Lloyd and Mauldon (1986) and for India, Crawford (1977)).

In the United States, Senate hearings were held on the adequacy of private crop insurance plans as early as 1923. After the droughts of 1934 and 1936, the debate on crop insurance was renewed, this time as part of the presidential election of 1936. President Franklin D. Roosevelt supported a government-sponsored crop insurance program, whereas his opponent Governor Landon preferred privately underwritten crop insurance. After Roosevelt won the election, Senator James P. Pope of Idaho sponsored the Federal Crop Insurance Act, passed in 1938 as part of the farm bill (Kramer 1983).

The U.S. crop insurance program was set up as yield insurance. The same basic program exists today. As yield insurance, indemnities are paid when a farmer's actual yield falls below some predetermined threshold yield, and indemnities are the shortfall in yield below that threshold multiplied by some predetermined price. The program has changed over time in the geographic areas and crops it covers, in the subsidy attached to premiums, in the calculation of the threshold yields, in the method of payment of premiums and indemnities, and in the timing and length of contracts.

How did the U.S. program perform? The Federal Crop Insurance Corporation (FCIC) suffered continued losses in its first years, and was cancelled for more than a year in 1943. It was revived on a restricted experimental basis in 1945. Loss ratios were above one for each crop covered till 1945; the loss ratio for all crops together fell below one for the first time in 1947. Historically, a loss ratio of unity or less has been generally considered satisfactory, indicating approval of public assumption of the costs of program administration. However, the program was severely curtailed in 1948 in terms of the counties it covered. The years 1948 to 1952 saw an overall FCIC loss ratio below one while many farmers were protected from bankruptcy in the drought of 1951 and 1952, but the program failed to accumulate reserves. The late 1950s also brought good loss ratio years for the FCIC, but expansions in the 1960s eroded the financial position of the FCIC. The passage of bills in 1973, 1975 and 1977 to create and renew a disaster relief program was believed to have affected the financial performance of the crop insurance program. The availability of such a zero-cost risk management instrument reduced the attractiveness of purchasing all-risk coverage.

In an effort to make crop insurance the primary form of disaster assistance, the Federal Crop Insurance Act of 1980 was passed, greatly expanding the counties and crops covered, enabling private companies to offer insurance coverage with FCIC reinsurance, and directly subsidizing premium payments up to 30%. But the aim of eliminating ad hoc relief measures was not achieved; bills to provide disaster aid to farmers were passed in 1983, 1986, 1987, 1988 and 1989. The percent of eligible acreage covered increased steadily through the 1980s (except for 1983, the initial year of the Payment in Kind program), rising from 9.6 percent in 1980 to 24.5 percent in 1988 (U.S. General Accounting Office 1989). (Although participation jumped to roughly 40 percent in 1989, the increase occurred because farmers who collected disaster relief in 1988 were required to purchase insurance in 1989.) Total premiums paid by farmers followed the same upward trend during the 1980s, but so did indemnities. Calculating the loss ratio as indemnities divided by premiums paid by farmers (excluding government studies), we find that the crop insurance program has not improved, with loss ratios ranging from 1.23 to 3.46. The FCIC reports the loss ratio (including subsidies in the premiums) for the 1980s as 1.58. This figure rises to roughly two when the premium subsidy is removed, and to roughly 2.5 when the administrative costs of running the FCIC are included. That is, more of the costs of the program, on average, are paid from the public purse than from farmers' contributions.

As an indication of the overall success of this oldest of the world's public crop insurance schemes, we note that the FCIC loss ratio, in almost fifty years of experience, has been below one just nineteen times, the lowest ten observed values being 0.26, 0.50, 0.53, 0.58, 0.60 (three years), 0.65, 0.68 and 0.69 (U.S. Department of Agriculture, Federal Crop Insurance Corporation, 1981; U.S. General Accounting Office 1989). Thus the twentieth percentile loss ratio is uncomfortably close to a value which should be well above the median, given the skewed distribution of indemnities.

In Japan, public agricultural insurance also began in 1939, initially structured as rent insurance. Insufficient coverage led to the adoption of a yield insurance program in 1947. The chief differences between the Japanese and U.S. plans are that insurance is compulsory (for farms over about 3/4 acres) and that coverage is available on a plot basis. Like the U.S. program, premiums are heavily subsidized (over 50%) and administrative costs are also subsidized (Yamauchi 1986; Crawford 1977). It is clear that any reasonable cost-benefit analysis would show that this program, which encourages, at substantial cost, an increase in already excessive domestic rice production, with regressive effects on income distribution, is a socially wasteful use of resources (Roberts, Gudger and Gilboa 1989).

In Sweden, a compulsory crop insurance program has been in effect since 1961 (Crawford 1977). Premiums are paid as levies on farm deliveries, but the deductible on each farm is adjusted by the variability and number of crops grown. The government also contributes a subsidy of more than double the farmers' contribution. The loss ratio net of subsidy was 1.78 for the period 1961 to 1971. Despite this large ratio of indemnities to contributions, many farmers are dissatisfied with the program since it does not always indemnify their individual losses, indemnification being determined on a district-wide basis.

Brazil has experimented with a production cost insurance, a form that was also tried experimentally in the 1940s in the United States (Lopes and Dias 1986; Crawford 1977). Coverage was based on out-of-pocket production expenses. Payment of indemnity occurs when yields are too low to cover the costs of production, and is equal to the shortfall. In the states of Sao Paulo and Minas Gerais, state-owned insurance companies (COSESP and COSEMIG, respectively) offer coverage, with participation mandatory for cotton farmers in Sao Paulo, voluntary for other crops and for Minas Gerais. Although the livestock program had loss ratios below one, leading to a premium reduction, and the cotton program had a loss ratio of close to unity for the first eleven years, the COSESP has had a loss ratio for all crops of 1.33 from 1971 to 1980. In Minas Gerais, participation is very low, and the loss ratio for 1973 to 1979 was 3.05, for all crops.

A related form of agricultural insurance is credit insurance. Coverage is based on the amount of credit extended to a farmer, and the insurance is mandatory for access to official credit. Indemnities are equal to loan repayments less revenue from harvested crop. This type of insurance has been used in Brazil (PROAGRO), with disastrous budgetary consequences (Lopes and Dias 1986; Crawford 1977). The loss ratio for this program ranged from 2.44 to 4.2 from 1975 to 1981, with an overall ratio of 3.87

(Lopes and Dias 1986). Cotton farmers in Sao Paulo with official credit prefer the COSESP insurance to the PROAGRO insurance despite higher premiums, because early-season damage is quickly indemnified and replanting is possible. In 1980, nearly two-thirds of the costs of the PROAGRO program had to be met by the federal budget and the central bank, with severe macroeconomic consequences.

Another country with a government subsidized credit insurance plan is Panama (Pomareda 1986; Hazell, Bassoco and Arcia 1986). This plan had a loss ratio of less than one from 1976 to 1981, but large losses in 1982 increased the loss ratio for the whole period to roughly 3. (This example emphasizes the general point that skewness of results makes a short sample of loss ratios an unreliable indicator of long-run viability.)

Mexico began an all-risk credit insurance program in 1964, which is compulsory for official credit (Bassoco, Cartas and Norton 1986; Hazell, Bassoco and Arcia 1986; Crawford 1977). Many private lenders also require insurance as a loan condition. One effect of this has been to limit farmers' use of credit. A 1973 survey showed that 38 percent of small farmers in Puebla did not know what crop insurance was. Intercropping of beans and corn, common among the poor in rural Mexico, is not insurable, but is itself a risk-reduction tool. The insurance program had a loss ratio of 0.89 from 1956 to 1976, but this ratio includes the governmental subsidies. Indemnities mounted through the 1980s, to the point where the current government, beset by budget pressures, recently proposed dropping the program.

In the 1970s, the Philippines had a credit insurance type scheme, where a 2%-of-loan premium contributed to a fund guaranteeing loans. This guarantee only postponed a farmer's loan payments in bad times. Thus, many farmers fell in arrears due to successive bad years. The Philippine Crop Insurance Corporation replaced this scheme in 1981, mandating production cost insurance for farmers with official credit. This subsidized scheme performed well through 1982, but has seen loss ratios above one (including the subsidy of premiums) in the years 1983 through 1985 (Muyco

1987). On the other hand, a mutual aid fund for credit insurance was piloted in one village in 1983. Farmers control the fund: they may withdraw with their contributions at any time and voted to increase the premium from 1% of their loan amount to 3%. During an 18 month period in 1984 and 1985, the farmers had a 100% repayment rate, allowing the fund balance to accumulate (Technical Board for Agricultural Credit 1987).

Sri Lanka's crop insurance scheme has been in operation since 1958, when a three year pilot program began. Since its inception, the program has been meant to cover average or normal losses beyond a farmer's control, using collected premiums. Abnormal or catastrophic losses are to be covered by the government as relief. Although insurance became mandatory for paddy rice in certain areas in 1973, less than 40% of premiums were collected, since farmers were allowed to pay the premiums after the harvest, that is, after observing whether they sustained a loss. The loss ratio for 1961 to 1973 was observed to be 2.45, based on collected premiums only. In 1973, the program was expanded to cover the entire country, and the Agricultural Insurance Board was given the authority to prosecute farmers in default. But the enforcement provisions were rarely used and loss ratios below one prevailed in only six of sixteen growing seasons. The highest loss ratio for the period (1975 through 1983) was only 2.7, but this is because indemnities were scaled back when the insurance fund (plus any extra government contributions) was insufficient to make full payments. Changes in 1983 expanded the crops covered, removed the government option of prorating indemnity claims, and instituted an incentive to make no claims, allowing premium-free participation to farmers with five consecutive no-claim seasons. The seasons for which data is available had loss ratios in the range 0.28 to 5.14 for all crops, and 1.16 to 2.07 for rice (Amarabandu 1987).

In Mauritius, the compulsory program for sugarcane was begun at the behest of the British government, after issuing disaster relief in 1945 (Crawford 1977). Cyclones and drought are the chief hazards in Mauritius. Although the insurance fund ran a surplus until 1959, losses in 1960 wiped out the entire fund. The overall loss ratio for 1957 to 1973 was 1.05. This program is a disaster insurance program run much as a commercial scheme would be. Loss is determined by the government declaring the whole island or some portion a disaster area—farmers not included may appeal to the Supreme Court. Thus losses are very low in most years and very high in just a few years. (This is the pattern seen for the U.S. disaster payments, shown in Fig. 1 below, in contrast to regular all-risk crop insurance for individual-farm losses.) In normal years a balance is maintained in the reserve fund which is sufficient to earn income greater than the administrative costs (Roberts, Gudger, and Gilboa 1989). The administrative costs are low, due to the compulsory nature of the scheme. When the loss ratio (for the period 1969 to 1971) is distinguished by size of farm, it becomes obvious that income transfer is one effect of the insurance program. The loss ratio for small farmers is 1.06, while it is 0.47 for large farms. The fund suffered as indemnities were paid in the years 1981 through 1984. However, the fund was not erased and a target level at which to maintain the fund has been established.

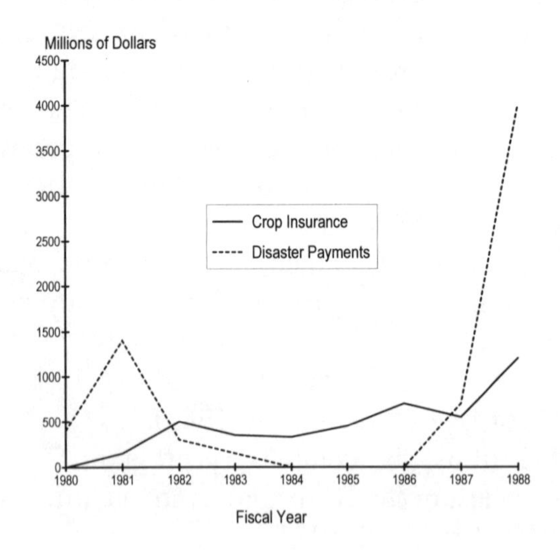

Fig. 1. United States crop insurance and disaster payments

Source: U.S. General Accounting Office 1989

India began a nationwide crop credit insurance program in 1985 (Sen 1987), after several serious attempts to institute a program, beginning in 1947, failed (Crawford 1977). Although it is too early to gauge the success of this program, its performance may well be enhanced as a result of the pilot program and study which preceded the full program. However, the premiums were cut by over 50% in moving to the full program, reducing the likelihood of sustaining the loss ratio (including subsidies) of roughly unity for the 1979 to 1984 pilot period.

The prudence of assessing the need for, and feasibility of, all-risk crop insurance before committing resources to such schemes, seems elementary. In Australia, several studies of farmers' demand for crop (or rainfall) insurance have shown that such programs would not attract sufficient participation for viability, despite the extreme environmental uncertainty experienced by Australian farmers (Bardsley, Abey and Davenport 1984; Industries Assistance Commission 1985, 1986; Patrick, Lloyd and Cary 1985; Patrick 1988). Fire and hail insurance is provided privately, and, as in the United States and Canada, is very widely used where needed.

Many other countries have explored the possibility of instituting crop insurance schemes (Crawford 1977). Some countries did institute plans, though they were restricted to specific risks such as fire or hail (Cypress, Greece, Argentina and Puerto Rico are examples). Other countries have commissioned studies, and even recommended the institution of a crop insurance plan, but for whatever reasons, have never done so (for example, Columbia, El Salvador, the Philippines, Malaysia, and Turkey). Israel has a crop insurance program which compensates growers for production costs (60 percent in the central and northern regions, 100 percent in the south) whenever yield falls below a break-even level. Recent reports imply that the expense of the scheme is prompting calls for its alteration or abolition. South Africa's program is not subsidized and had a loss ratio of 0.68 for 1970 to 1976, but is primarily a hail insurance scheme.

Our brief overview of the extensive history of all-risk crop insurance and similar programs in a large and diverse set of countries leads us to affirm the conclusion of Hazell, Pomareda and Valdes (1986, p. 296): "Overall, the findings of this book are not encouraging for crop insurance, and governments would be well advised to look carefully before embarking on large and costly multiple-risk crop-insurance programs". This conclusion is echoed in Roberts, Gudger and Gilboa (1989), who cite trends towards specific risk or named peril insurance and commercially offered insurance.

4.2 The Standard Theoretical Approach

Why has the history of all-risk crop insurance been so dismal? Has the design of the contracts been insufficiently clever? Has the execution been unusually incompetent or corrupt? Or is there another, more general, explanation? Consider the following: "In theory, agricultural insurance is an efficient risk-sharing mechanism. In practice, however, agricultural insurance has been a costly method of transferring risk from farmers to governments and/or other insurers" (Nelson and Loehman 1984, p. 523).

The above introduction to a standard reference on agricultural insurance typifies the view of works on agricultural insurance theory. In that view, the theory is fine, the application just needs a little modification. "...[P]ublic subsidization may not be necessary. Information collection and application of contract design principles are possible ways of achieving the benefits of insurance at less cost than public subsidies" (Nelson and Loehman 1984, p. 531). Further elaboration of insights from the mechanism design literature are found in Chambers (1989) and chapter 5 of this book.

The theory to which these writers refer is good theory, but it does not apply to the crop insurance problem. As a basis for the analysis of crop insurance, it has several deficiencies, as some writers have long since recognized (for example, Roumasset 1978).

First, the program for which it is relevant is not crop insurance but insurance of total revenue. Moreover, alternative means of handling risk that are observed to be important in practice, and are affected by insurance, are absent from the analysis.

In the standard theoretical approach to agricultural insurance (here we follow Chambers 1989) the farmer is modeled as producing a vector of outputs y from inputs x, given random production disturbances (weather, pests) denoted by $\tilde{\theta}$ with joint density $h(\theta)$. Thus realized output is $y = y(x, \tilde{\theta})$, where θ is a particular set of realizations of $\tilde{\theta}$. The outputs y are sold at [illegible] prices, $\tilde{p}$, where the joint density of $\tilde{p}$ is $v(p)$. Total [illegible] indemnities $I(R)$ are added to, and insurance costs δ [illegible]ues to obtain the farmer's realized net income π, which is thus [illegible]d as

$$\pi = R + I(R) - \delta - wx, \qquad (1)$$

where w denotes input prices (assumed nonrandom). Th[illegible]rmer maximizes expected utility, defined (for example, Chamber[illegible]89) as a function of net agricultural income π:

$$EU = \int_{R_0}^{R^0} U(R + I(R) - \delta - wx)\, d\, G(R, x), \qquad [illegible])$$

where $G(R, x)$ is the cumulative conditional distribution of R given x, implied by $y(x, \theta)$, $h(\theta)$ and $v(p)$, and R_0 and R^0 are minimum and maximum possible levels of revenue.

The above simple representation of the objective focuses the analytical exposition on a problem upon which great progress has been made in the last two decades: the design by an expected-profit-maximizing insurer of an insurance contract for a farmer with this type of objective function and no other means of adjusting to risk. A fundamental issue is "insurability": Under what conditions does a policy exist that makes farmers and insurers better off than in a world with no risk-reducing instrument?

"Insurability" is an issue because the lack of a competitive market for insurance is a perceived policy problem. Theoretical attention is directed at the two classes of market failures: adverse selection and moral hazard. Incomplete observability is a feature of both problems. Adverse selection arises in crop insurance when differences in the exogenous riskiness of different farmers' crops are not fully observed by the insurer. (For a demonstration of the empirical relevance of this problem, see Skees and Reed 1986.) If premiums are set to yield non-negative profits for the insurer on the average-risk crop contract, low-risk farmers' crops will tend to go uninsured, unless coverage is mandatory, increasing the average riskiness, and making the contract unprofitable for the insurer. There may be no equilibrium in the absence of information helping to distinguish between risk classes (Rothschild and Stiglitz 1976; Ahsan, Ali and Kurian 1982).

In practice this problem may not preclude all private crop insurance. Crop insurance sellers (for example, sellers of hail insurance) may use regional or individual histories on yields, or on relevant weather information, to help separate risk classes so that it is feasible to offer different contracts to farmers with different degrees of risk.

Moral hazard can be defined as "[A]ctions of economic agents in maximizing their own utility to the detriment of others, in situations where they do not bear the full consequences or, equivalently, do not enjoy the full benefits of their actions *due to uncertainty and incomplete or restricted contracts* which prevent the assignment of *full* damages (benefits) to the agent responsible" (Kotowitz 1987, p. 549; emphasis as in original). With respect to crop insurance it has been discussed more narrowly in terms of input choice (Nelson and Loehman 1984; Chambers 1989; Glauber, Harwood and Miranda 1989) and when the adverse effects are addressed at all, they are described as increases in the *probability of collecting* indemnity payments (Glauber, Harwood and Miranda 1989; Nelson and Loehman 1984; U.S. Congress 1989; Ahsan, Ali and Kurian 1982), rather than in the size of the payments. In line with this viewpoint, measures for the control of moral hazard in crop insurance tend to focus on the monitoring of qualitative input choices (for example, cultivation, irrigation, and planting timing) on the one hand, and design of the contract on the other. For example, the existence of a deductible is justified as a means of mitigating moral hazard.

The theoretical literature on these issues does not go far in distinguishing agricultural crop insurance from other common types of insurance (automobiles, buildings, and health, for example). There are few good insights on the reasons why competitive private all-risk crop insurance has been virtually a universal failure, while other types of private insurance have succeeded (including, for farmers, hail insurance). We suggest that if we are to make headway answering this question, we must broaden the scope of the analysis.

Why should adverse selection and moral hazard, as described in the literature, be more serious for crop insurance than for health insurance or property insurance? It is not clear that they are, given effective administration. A report of a sample review of United States crop insurance by the U.S. General Accounting Office (November 1987) shows many egregious errors, heavily biased toward overpayments of indemnities. But the report indicates that most of the large errors detected by the U.S. General Accounting Office were in the administration of sales and/or adjustment of claims by private reinsured companies. The examples given do not involve subtle errors of observation or inference, but rather such gross errors as misclassification of unirrigated land as irrigated, or of land uncultivated by the required date as already planted.

Managerial responses of farmers that were not proscribed by law, or were undetected by the U.S. General Accounting Office, may have had important social costs. But in the contexts of the cases discussed by the U.S. General Accounting Office, we think it reasonable to infer that the major source of moral hazard, as conventionally interpreted (increased probability of crop loss) stems from the dissociation of performance from remuneration among the private corporations who as agents handled a greatly increased share of the retail business in the public crop insurance contracts over the latter part of the 1980s. That is, the policemen of moral hazard were themselves afflicted with the problem. (These problems persist. In 1990, it was reported that California Southern San Joaquin Valley farmers could collect as much as $20 million on insurance of unirrigated safflower, when a policy designed for higher-rainfall regions was mistakenly marketed to them via private agents (Schacht 1990).) Contracts directly handled by master marketers under the direct control of the public insurer (FCIC) appear to have had far less serious errors. Competitive private insurers would be even less likely to have these administrative problems.

Thus the conventional explanations in terms of moral hazard and adverse selection are unpersuasive. So the question remains: wherein lies the true explanation of the failure of competitive multi-peril crop insurance? The short answer is simple. Multi-peril crop insurance is worth less than what it costs, if full costs are covered by premiums in the long run. The reason this conclusion is missing in the theoretical discussions is paradoxical: to get such a simple answer, one must use a more sophisticated, realistic and comprehensive model of the value of all-risk crop insurance on the farmer's wellbeing. In such a model, the cost of uninsured risks, and the value of revenue insurance, are much less than conventional approaches imply. One major reason for this is that there are endogenous responses to agricultural risks not visible in the standard depiction of moral hazard, and crop insurance is to some unquantified extent a substitute for these. Roumasset (1978, p. 93) recognized this, when he stated that "[T]he best we can hope for by instituting a crop insurance program is that risk premiums will become slightly smaller, and as a result, farmers will be slightly less risk averse." Furthermore, revenue insurance is not constrained optimal in the standard model unless it is identical to income insurance. But, it is not identical if some input prices are random when production commitments are made. Crop insurance is generally less efficient than revenue insurance.

and sheep are very frequently produced together, and in many less developed economies multiple crops and inter-cropping are common, including agro-forestry (see Walker and Jodha 1986).

The possibility of crop diversification complicates the optimal rate structure as conventionally analysed. An independent deductible for each crop, for example, does not have the same incentive effects as a deductible for the whole farm's income protection. If crop A is failing, for example, it may pay the farmer to concentrate his effort on crop B, and increase the insured loss on crop A. If, under the same rate structure, the farmer were producing only crop A, he might find it preferable to work harder to prevent the deductible loss. The design of crop insurance contracts, in practice, does not take account of these possibilities.

Diversification extending beyond the farm operating unit is ignored in the crop insurance literature. First, earnings of the farm family's labor endowment can be diversified. Off-farm work is an important source of income in many countries, and in developed countries including the United States, the European Community, and Japan it is likely to be non-agricultural work, with income not highly correlated with farm income. In less developed countries an important source of off-farm income may be remittances from family members who have migrated to find work.

Diversification of sources of capital is also ignored in analyses of contract design. The objective in (2) assumes no external financing. Yet practical discussions emphasize that crop insurance should be most attractive and important for farmers with heavy debt burdens. Certainly these are the farms in greatest danger of failure. There is room for a neater marriage of theory to observation here. As Innes (1988 and 1990) has pointed out, there are good reasons (moral hazard and adverse selection) why off-farm financing comes in the form of debt rather than equity, given that equity would more uniformly share exogenous risks with the farmer. But given the use of debt, the farmer's attitude toward risks of large losses. At the margin, losses in excess of the level that forces foreclosure are born by the lender. This means a farmer's behavior may be in effect risk-loving when the financial situation is precarious. If foreclosure is certain even when output is as high as "normal", chances of survival may be increased by decisions that increase the variance of revenue. For example, refusing insurance can in some circumstances increase the probability of staying in business.

The U.S. Congress (1989, p. 61) reports that "Insured producers tend to purchase too much insurance for relatively common events and too little insurance for low probability events that are beyond their financial capacity. Quite simply, people prefer to insure for small probable events rather than large improbable events." The rationale offered by the U.S. Congress (1989, p. 60) is that "[I]ndividuals tend not to think about events that have a low probability of occurring, even though these events may have potentially disastrous economic consequences." This statement is inconsistent with expected utility analysis, though empirical evidence is more mixed. Some experimental studies imply that low-probability, large-loss events tend to be over-weighted by individuals, rather than ignored, though it is reported that evidence from experience with flood insurance shows the opposite (Kunreuther, Sanderson and Vetschera 1985). But the farmers' attitudes in favor of insurance for small losses make sense (and may be consistent with expected utility theory) if some of the extra costs of larger catastrophes are borne by their creditors. In such circumstances the creditors share the benefits from insurance of large losses without compensating the farmer for the purchase of this insurance.

4.4.2 Intertemporal Smoothing. In (2), there is no time dimension. No lag is recognized between input decisions and output realizations. If inputs are chosen as outputs are produced, the relevant state of the world (for example, the weather) is known as inputs are chosen. That is, the production function has no risk at the time when inputs are committed. Risk aversion does not affect allocative efficiency at all. Of course this interpretation is too literal. Modelers who use an objective like (2) have in mind a situation where inputs are chosen before outputs are produced and the relevant state of the world is revealed. Chambers, for example, models input prices as deterministic but output prices as random, just the situation at decision time if a lag occurs between input commitment and output, as in actual agricultural production.

In the context of farmers' insurance decisions, the fact that the objective (2) is static, acknowledging no intertemporal linkages is perhaps the greatest defect of the standard theoretical approach. In such a model, net income equals consumption; there is no saving or investment. But, as already noted, inputs (for example, seeds) must be used some time before the harvest. Any subsistence farmer who plants seeds or stores them, instead of eating them, is saving or borrowing and investing at the expense of lower current consumption. To operate at all, farmers must be forward-looking and make choices about intertemporal reallocation of consumption. They are able, if they so desire, to adjust their saving/borrowing to cushion shortfalls and smooth out unexpected increases in consumption.

Such financial smoothing is an important feature of risk management, in modern finance theory and in practical agriculture. Consider, for example, farmers in the Australian Mallee, where the coefficient of variation of wheat yields for a sample of sixty farms is estimated to be 41 percent per year with no trend (Patrick 1988). "Farmers were asked open-ended questions about actions they would take to: (a) reduce the magnitude of and (b) cope with the impacts of year-to-year income variability.

"Financial responses were mentioned more than twice as frequently as production responses. Almost all farmers indicated that they built reserves or reduced debts in good years. Cash flow planning was reflected both in making investments in a good year as well as reducing family and farm expenses" (Patrick, Lloyd and Cary 1985, pp. 21-22).

The use of savings to smooth consumption in the face of income fluctuations has been recognized by general theorists (if not risk analysts) at least since the seminal work of Friedman (1957). (See Deaton 1991 for a modern reference.) A recent study of eighteen Illinois farmers reports the short run marginal propensity to consume at less than 0.02 (Langemeier and Patrick 1990). There is no reason to expect such smoothing behavior to be confined to countries with well-developed financial markets. Paxson (1988), in a careful econometric study of responses to income shocks in Thailand, concludes that her results "provide a needed link in the hypothesis that declines in Thai savings since 1981 are partly due to unanticipated declines in farm incomes. Specifically, the results suggest that farm households do respond to unanticipated declines in farm income by reducing savings close to baht for baht" (p. 41).

Walker and Jodha (1986) also show the importance of distinguishing consumption from income. Results over five cropping years for three villages in a semi-arid area of India show an average coefficient of variation of income per person of 35 percent, with a range of 15 to 85 percent. Between drought and nondrought years, the *smallest* falls in crop and livestock income were 58 percent and 37 percent. The *largest* drop in household consumption expenditure was 12 percent (with a range of only 8-12 percent) and the largest drop in foodgrain consumption was 23 percent. Private borrowing was a major source of smoothing, contributing 44 to 73 percent to private "income" (sic) in the drought year, more than public relief. Outstanding debt increased a minimum of 64 percent.

Once intertemporal reallocations of resources are recognized, the specification of utility in (2) becomes inappropriate. In all modern theories of utility, it is consumption not income that is the argument of the utility function. Risk affects utility via consumption fluctuations, and insurance is valuable to consumers only insofar as it reduces consumption fluctuations in a way that, when insurance costs are taken into account, is preferred by them. But in (2) above the argument of the utility function is income, not consumption. Even if the concept is enlarged to recognize diversification responses and to include off-farm income, it is the wrong variable for welfare evaluation. This point may seem elementary, but is regularly missed in the agricultural insurance literature.

No one expects complete smoothing of consumption in the face of large income fluctuations, and a twelve percent cut for poor Indian farmers in the example above must have been a harsh burden. But the marginal value to a risk averter of reducing very large consumption shortfalls decreases with the magnitude of the reduction. For example, moving each possible outcome half way to the mean without changing its probability will reduce the cost of risk by roughly three quarters, for "small" risks or quadratic utility. Existing alternatives for partial smoothing make the marginal contribution of crop insurance less valuable than is implied, at any level of risk aversion, by standard analyses equating consumption to income. If insurance offered on competitive terms (recognizing the true opportunity cost of resources) by private or public insurers is accepted by farmers, then such insurance may be socially valuable, unless it exacerbates other sources of market failure such as environmental externalities. Its value is not, however, the full value (net of costs) of the smoothing of consumption achieved by the insurance, but the value relative to what could be achieved in the absence of the insurance. When farmers don't buy insurance on competitive terms, the simplest inference is that this value is negative. A subsidy can make crop insurance a net burden on society, relative to other ways of giving farmers benefits equal to any received from the insurance. Before considering the neglected sources of this net burden, let us pause to consider just who receives any of the "farmer benefits" from the program.

4.5 The Incidence of Crop Insurance

Failure to recognize the intertemporal nature of the problem not only leads to overestimation of the benefits of insurance through neglect of important private risk management strategies, but can also lead to a mistaken view of who gets the benefits that are provided. To the extent that insurance increases output of food and fiber, as anticipated in the introductory quote, it is likely to reduce output prices and, if demand is inelastic, producer revenues. Consumers gain from lower prices, but they lose as taxpayers.

Let us assume, for now, that introduction of all-risk insurance does not increase output but does increase the utility of the discounted expected flow of consumption of current farmers. Surely this type of insurance is a contribution to the continued wellbeing of farmers and the protection of farming as a way of life? Not necessarily. If insurance makes a farm operation more attractive to future farmers, they will compete to bid up the purchase price, till in a competitive land market the winning bidder is about indifferent between farming and the next most attractive opportunity.

This land-price effect is not just a theoretical implication of competition. It is detected empirically as capitalization of disaster payments in risky U.S. regions without crop insurance by Gardner and Kramer (1986). The bulk of the gain goes to those owning the land when the price is bid up because of the introduction of the program, whether or not they stay in farming (Myers 1988). The winning bidder, if particularly risk averse, may gain a little from the insurance, but bidders less risk averse than average may be actually worse off than without it.

4.6 Moral Hazard—A Broader View

As mentioned above, moral hazard has been defined in the literature as implying an increased probability of loss or receipt of indemnities. In the context of crop insurance, the definitions noted earlier must be broader. It is well recognized that crop insurance can change not just the probability of a crop loss, but the size of the loss per acre (through input intensity), and the number of acres affected (through changes in land use), and even, by its effect on sod-busting of marginal land, the integrity of the environment (Gardner and Kramer 1986). But crop insurance also changes the use of other risk management measures including those mentioned above: spatial and activity diversification, off-farm employment, use of implicit or explicit contractual relationships, savings and/or loans for consumption smoothing, and so on. The responses to other policy measures noted by Walker and Jodha (1986, p. 24) might well apply to crop insurance: "In India many well-intentioned public policies have generated side effects that have made risk management by small-scale farmers less effective in drought-prone areas. Intrayear reserves and intrayear security stocks of food grains and fodder have ceased to be important components in risk adjustment [Jodha 1981]. Group measures such as mutual risk-sharing arrangements, seasonal migration, and informal interlocking of agricultural factor markets are less compatible with new village institutions. Legal provisions regulating credit, labor contracts, mortgage of assets, and tenancy are often insensitive to the specific adjustment problems of drought-prone areas [Jodha 1981]. For these and other reasons, formal public relief has assumed greater significance in drought-prone areas."

The inescapable reality is that all-risk crop insurance has been shown repeatedly, in a large variety of circumstances, to require heavy subsidies to obtain widespread acceptance. If, as the subsidies are increased, farmers move from more efficient means of risk management into crop insurance, this move causes a social loss. The subsidy on crop insurance thus induces moral hazard in risk management. And a major part of the cost may be in financial management, rather than production management which has been the dominant focus of attention. All-risk crop insurance has high administrative costs and characteristic managerial difficulties (see, for example, U.S. General Accounting Office November 1987, and the diverse examples of ways to lose control of crop insurance programs presented in Hazell, Pomareda and Valdes 1986). It is a socially costly substitute for alternative (admittedly imperfect) private methods of managing farm yield risk.

Farmers and policy-makers from the United States and Canada might agree that a viable public all-risk crop insurance scheme is costly, but might believe, as the quote at the start of our paper implies, that public assistance with risk management is necessary for a viable competitive farming industry. One of the pernicious aspects of experience with such assistance is that it tends to reinforce such beliefs, so that the assistance can be self-justifying. Subsidized crop insurance, if widely accepted, drives out alternative coping mechanisms, and encourages responses that increase risk exposure.

For example, one response to highly subsidized crop insurance might be to demand more debt. Banks may well be willing to comply. Just such a response is envisaged, with apparent approval, in the quote at the start of the paper. This adjustment increases the riskiness of the farmer's equity position, counteracting the "stabilizing" effects of the insurance, but making the insurance seem more indispensable.

The effect is analogous to the impact of a public storage scheme on private storage (Peck 1977; Wright and Williams 1982; Williams and Wright 1991). Public storage induces compensating reductions in private speculative stocks. Then the inadequacy of private stocks can be used as a justification for continuing the public scheme. Similar responses may be elicited by price support schemes. Farmers may also become more specialized, less flexible in their managerial capacity and the design of their farm operation. In this new situation the necessity for public subsidies and support attains great plausibility.

It is not widely recognized by policymakers and economists that farmers can operate efficiently with low failure rates in very risky environments, with very little public support. Those 60 Australian Mallee farmers mentioned above averaged 45.5 years of age, and most had been farmers since leaving secondary school, and on average had lived on the same farm since before that time (Patrick, Lloyd and Cary 1985). Of those with children, 84 percent anticipated their children would go into farming. Their next-to-last crop had been obliterated by drought, and 73 percent of the farms had gross income of less than $A 25,000 that year. Yet after one more crop year, only 7 percent of the sample had farms whose net worth was less than $A 250,000, and almost one quarter had net worth above $A 1,000,000. Only 7 percent believed they would not survive one year with half the average yield; more than half believed they could survive five consecutive years with such meager yields.

A strong equity position was obviously a crucial factor in such durability. Modal net worth was in the range of $A 500,000 - 749,999 but modal debt was only $A 50,000 - 99,999, with about one-third having no debt at all (Patrick, Lloyd and Cary 1985). Farmers also diversified; more than 90 percent had breeding ewes, and more than 90 percent grew barley as well as wheat.

Despite the extreme annual risks they face, the majority of the sample was not interested in hypothetical actuarially fair multiple peril crop insurance or rainfall insurance. But the farmers were not ignorant of, or averse to, insurance in general. More educated farmers were not more willing to adopt the insurance packages. And fire and hail insurance, private health insurance, and supplemental automobile insurance in addition to compulsory coverage were each held by more than 90 percent of the sample (Patrick, Lloyd and Cary 1985).

The durability in the face of, by North American standards, extreme yield risk (not to mention the additional price risk) seen in this sample is replicated throughout the Australian wheat and sheep industries. Though the 1980s have been considered a very difficult time for farmers in Australia, the failure rate has been only one or two percent. Yet government assistance to the wheat and sheep farmers has not been very significant, and after some deliberation (Bardsley, Abey and Davenport 1984; Industries Assistance Commission 1985, 1986), public crop insurance has been rejected as a policy option. There is no evidence that lack of such a program has handicapped producers in their ability to compete effectively in world markets.

The reluctance of the Mallee farmers to consider rainfall insurance is noteworthy. This type of insurance can be arranged in effect as a "weather lottery", and as such should be free of adverse selection and moral hazard on the part of the putative purchasers. And farmers know rainfall is important: 90 percent of farmers ranked the quantity and timing of rainfall as the number one source of variability. The conventional theoretical explanations for failure of insurance markets—adverse selection and moral hazard—do not seem to explain the failure of this type of insurance to attract greater interest. One would imagine that administrative costs of such a scheme would be lower than for conventional insurance. The experiment suggests that farmers do not find existing risk management strategies as inadequate as many economists and policymakers appear to believe.

4.7 A Superior Substitute for Disaster Payments?

In recent discussion of all-risk crop insurance in the United States, the dominant justification for expanding the program is not improvement of the welfare of farmers or others relative to a world of no intervention. Rather, it is to eliminate the need for disaster payments. As expressed in the Federal Crop Insurance Commission Act of 1988, Sec. 2(a), "[T]he Federal insurance program was established to promote the national welfare by improving the economic stability of agriculture through a sound and comprehensive system of nationwide crop insurance that would allow the phasing out of disaster payments programs as part of farm price support and income protection programs" (U.S. Congress 1989, p. 135). The U.S. General Accounting Office (August 1988, p. 2) also supports insurance as "the best way to provide disaster assistance."

The need for disaster payments, absent crop insurance, is determined by political perceptions. During consideration of the Disaster Assistance Act of 1988, the question of lack of participation in the federal crop insurance program naturally arose: "After all, 1980 Congress completely overhauled the program to include coverage on virtually all crops in every county.

"However, the sad fact is that only about 25 percent of the acreage eligible for insurance coverage currently is enrolled in the program. This level of participation simply is not adequate enough to address the financial consequences of a disaster like we experienced this year. Nor is producer participation in the program sufficient to forestall the political pressures that always come to bear when natural disasters strike a wide path, especially in an election year." (Statement by Congressman Ed Jones during consideration of the Federal Crop Insurance Commission Act of 1988. See 134 Cong. Rec. H8230 (daily ed. Sept. 26, 1988).)

The judgement that substantial participation would be needed to allow termination of alternative sources of relief echoes a statement of then-Secretary of Agriculture Bergland in 1978: "If less than 60 to 70 percent of the farmers are protected, it is likely that a sense of sympathy will prompt the system to provide protection for those who did not participate. Reaching the target level of participation would require both a well developed program and the termination of other programs that would provide protection and thereby compete with the new system." (Statement by Hon. Bob Bergland, "Disaster Assistance for Farmers," Committee Print, Senate Committee on Agriculture, Nutrition, and Forestry, 95th Congress, 1st Sess. 112 (1978).) Note that Bergland introduces a second, related idea: termination of disaster assistance may be needed if crop insurance coverage is to reach its target.

We do not have the hubris to argue with seasoned politicians on the political substitutability of crop insurance for disaster assistance, although we note in passing that Bergland's apparent expectation that the crop insurance he proposed would become the sole disaster-protection instrument has not been realized. But the substitutability of the two programs from the farmer's viewpoint is questionable. A review by the U.S. General Accounting Office (July 1988) of four limited surveys of farmers cites expense as a major reason given for non-participation in crop insurance in all four surveys. The U.S. General Accounting Office review also notes that in a South Dakota State University survey, the primary reason perceived by 191 major agricultural lending institutions in South Dakota for farmer non-participation is that the cost exceeds the benefits. They only encourage their weaker clients to participate and the participation decision is price-sensitive. Another study of crop insurance in the General Accounting Office review, by the Texas Agricultural Policy Research Center for Texas cotton producers in selected regions, is even more negative: "The economic payoffs from multi-peril crop insurance for cotton are negative at all levels of coverage in the Southern Plains, the Coastal Bend, and the Mississippi Delta. This indicates the producer would be better off not to buy insurance at any level of coverage."

A crucial feature of the above responses is that the attraction of disaster relief as an alternative is nowhere mentioned by the farmers or bankers. (The influence of disaster payments is not necessarily implicit in the comments on expense or profitability as reported, since insured crops do not lose eligibility for disaster relief.)

In 1989, a national survey by the Economic Research Service of the United States Department of Agriculture asked farmers who did not have federal crop insurance to rank a list of specific reasons for non-participation. The number one reasons are listed in Fig. 2, by percentage of respondents. The top five

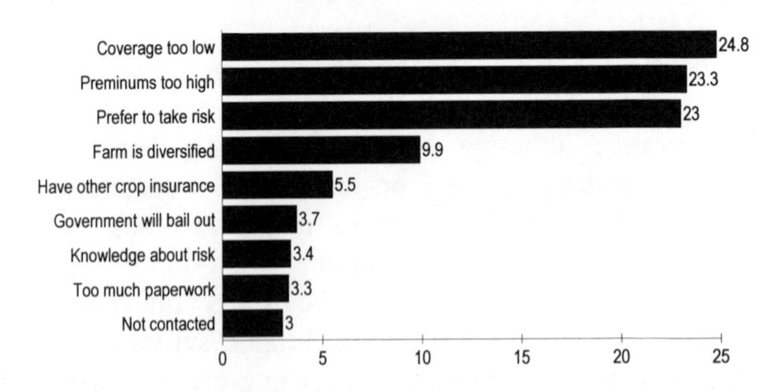

Fig. 2. Percent of farmers ranking the #1 reason they do not buy Federal crop insurance

Source: U.S. Congress 1989

reasons, covering almost ninety percent of respondents are really different versions of one general response: "The benefit wasn't worth the cost." Only 3.7 percent listed the anticipation of alternative government assistance as their principal reason. While 41 percent ranked this reason as at least "somewhat important," the rest ranked it as "unimportant" (U.S. Congress 1989). Given these responses, it is hard to argue that disaster assistance is the reason most non-participants do not buy crop insurance.

Other evidence against strong substitutability between crop insurance and disaster payments comes from the history of expenditures of the two programs since the inception of the new crop insurance program in 1980. Fig. 1 shows that public expenditures on crop insurance have been rather steady throughout the period; the loss ratio, calculated after the premium subsidy is added to the premiums, has been above unity every year between 1981 and 1988 (U.S. Congress 1989). On the other hand, disaster payments have two high spikes with relatively low expenditure in between. This difference is not surprising. Disaster payments occur only when a widespread threat to agricultural incomes materializes—a threat highly correlated across farms. Crop insurance pays out under a larger set of states of the world, including those where most other farms, *or even other parts of the same farm operation,* are doing well but the farm in question has suffered an idiosyncratic yield deficiency.

Paradoxically, another argument for nonparticipation is excessive optimism, which would make a program of disaster payments unattractive indeed: "Producers tend to forget bad years, and, as with most decisionmakers, they tend to ignore potentially disastrous events. In other words, individuals tend not to think about events that have a low probability of occurring, even though these events may have potentially disastrous economic consequences. Participation in the Federal flood insurance program has demonstrated this problem.

"...In interviews with 100 Kentucky grain producers, researchers found that producers tend to forget years when yields were low. The yields that over 60 percent of the corn and soybean producers indicated would be the very lowest yield they could expect in 1987 were higher than their farm records indicated their yields had been for the 1983 crop. Less than 10 percent of the corn and soybean producers indicated that yields could be lower than their 1983 crop yields" (U.S. Congress 1989, pp. 61-62). These observations are similar to those made by Kunreuther, Sanderson and Vetschera (1985) and mentioned above, but they are inconsistent with standard expected utility theory, and even less consistent with findings of many modern experiments on risk attitudes. Significantly they do not describe individuals who would be attracted by the alternative of disaster payments, but just the opposite.

This lack of correlation between disaster payments and crop insurance payments raises another problem. The political conjecture is that wider participation in crop insurance would sufficiently reduce the demand for disaster payments so that the latter could be cancelled. But it would seem to us that the constituency of disaster payments should include related businesses that would suffer from a deep and general regional cut in farm expenditures. These people might be much less interested in the added protection offered by crop insurance against individual income variations.

Finally, it is not so clear to us that ad hoc disaster payments are inferior in principle to crop insurance. The main objection seems to be that they are ad hoc and unpredictable. But disasters are unpredictable too. For farmers, the true catastrophes are multi-year droughts, depressions, and other persistent causes of very low income. Government should, and we hope will, act to protect farmers in these eventualities. But in such circumstances the assistance will inevitably be ad hoc. Insurance premiums are likely to be forgiven, and accumulated balances, if any, to be depleted, after a disastrous year or two. Already, under the Disaster Assistance Act of 1988, the requirement for disaster payment recipients in 1988 to purchase crop insurance in 1989 can be waived if the premium would exceed the county average by more than 25 percent, exceed the disaster payment received by 25 percent, or impose undue financial hardship (U.S. Congress 1989). If crop insurance were the only disaster assistance and large numbers of farmers claimed they could not afford the premiums, their coverage would be unlikely to cease. The emergency, as now, would be handled by ad hoc measures.

It is important, of course, that political decisions regarding disaster payments be well-informed. One elementary step in this direction would be to identify recipients of disaster payments by amount of payments and by taxable income level in the year payments are made, and make this information publicly available to taxpayers. If potential recipients need support, often far in excess of average income levels, at the public expense, such information seems a small price to pay in return. Unfortunately, this type of measure is, we have been told, unconstitutional in the United States. This difficulty would seem to merit further investigation.

At present, of course, the notion that premiums are expected to cover payments in the long run is something of a fiction. The term "actuarially sound" does not, for the FCIC, mean that the premiums cover costs, only that they cover expected indemnities. Even then, it has not been taken at all seriously. The report of the recent "Recommendations and Findings to Improve the Federal Crop Insurance Program" includes, in an Appendix by two members (U.S. Congress 1989, Appendix C, p. 111), the following comment: "The goal of actuarial soundness within 5 years is stated as an objective in the Commission report, but little specificity is provided as to how that objective should be accomplished." For a nation reeling from the steadily worsening budgetary implications of losses by federally insured Savings and Loans, these comments have a familiar but unpleasant ring.

4.8 Conclusion

In many different countries and circumstances, all-risk crop insurance has proven to be a failure, more wasteful of public funds and less attractive to farmers than the standard theoretical model would suggest. Many schemes may have been plagued by incompetence in design and execution, and by graft. But we do not believe that such factors are the major reasons for the overall disappointing results.

The main reason for the failure of all-risk insurance to meet expectations is that these expectations are explicitly or implicitly based on a theoretical model that overstates the potential value of such insurance, even if its administration were excellent and farmers are significantly risk averse with respect to consumption fluctuations. The problem with the model is that it is excessively simplistic. It ignores the many means for managing risk available to farmers in the absence of insurance, which greatly reduce the net additional effect of insurance in smoothing consumption, and make it unattractive to farmers when offered at a competitive market price. The explanations for market failure or "non-insurability" based on the standard model, adverse selection and moral hazard, are no doubt often important, but they do not tell the whole story. And they fail to explain why farmers seem no more interested in schemes that minimize these problems, such as rainfall insurance, than they are in crop insurance.

The argument currently popular in the United States, that some form of disaster assistance is politically inevitable, and crop insurance is the best form of disaster insurance, has fatal shortcomings. First, the two programs are not very close substitutes, as is evident from a comparison of expenditure flows on each scheme, and from farmers' own survey responses. Second, experience shows that an insurance scheme will not accumulate a fund sufficient to handle typical disasters. (In fact, in a normal year it is a much greater drain on funds than are disaster payments.) So either scheme would incur a drain on the public purse in such an eventuality. Third, experience after the 1980 Crop

Insurance bill showed that the government could not credibly commit itself to not providing extra funds in an emergency when the political pressure was there, even if a subsidized crop insurance scheme was in place.

The original political advocates of crop insurance recognized the paramount importance of economic and financial soundness. A half-century of experience with crop insurance programs gives a different, more negative twist to the opinion of Senator Pope of Idaho (1937): "The sociological need and background for Federal crop insurance is as strong and real as the economic and actuarial basis of the program."

References

Ahsan SM, Ali AGA, Kurian NJ (1982) Toward a theory of agricultural insurance. Amer J Agr Econ 64.3:520-529

Amarabandu WP (1987) Sri Lanka. Crop insurance in Asia. Asian Productivity Organization, Tokyo, 229-248

Asian Productivity Organization (1987) Crop insurance in Asia. Asian Productivity Organization, Tokyo, 267

Bardsley P, Abey A, Davenport S (1984) The economics of insuring crops against drought. Aust J Agr Econ 28.1:1-14

Bassoco LM, Cartas C, Norton RD (1986) Sectoral analysis of the benefits of subsidized insurance in Mexico. In: Hazell P, Pomareda C, Valdes A (eds) Crop insurance for agricultural development, issues and experience. Johns Hopkins Univ Press, Baltimore, Maryland, 126-142

Chambers RG (1989) Insurability and moral hazard in agricultural insurance markets. Amer J Agr Econ 71.3:604-616

Crawford PR (1977) Crop insurance in developing countries: a critical appraisal. Master's thesis, Univ Wisconsin-Madison

Deaton A (1991) Saving and liquidity constraints. Econometrica 59.5:1221-1248.

Friedman M (1957) A theory of the consumption function. NBER, New York

Gardner BL, Kramer RA (1986) Experience with crop insurance programs in the United States. In: Hazell, P, Pomareda C, Valdes A (eds) Crop insurance for agricultural development, issues and experience. Johns Hopkins Univ Press, Baltimore, Maryland, 195-222

Glauber JW, Harwood JL, Miranda MJ (1989) Federal crop insurance and the 1990 farm bill, an assessment of program options. Staff Report. AGES 89-45. USDA, ERS, Commodity Economics Division, Washington DC, 38

Hazell P, Bassoco LM, Arcia G (1986) A model for evaluating farmers' demand for insurance: applications in Mexico and Panama." In: Hazell P, Pomareda C, Valdes A (eds) Crop insurance for agricultural development, issues and experience. Johns Hopkins Univ Press, Baltimore, Maryland, 35-66

Hazell P, Pomareda C, Valdes A (eds) Crop insurance for agricultural development, issues and experience. Johns Hopkins Univ Press, Baltimore, Maryland, 322

Hazell P, Pomareda C, Valdes A (1986) Introduction. In: Hazell P, Pomareda C, Valdes A (eds) Crop insurance for agricultural development, issues and experience. Johns Hopkins Univ Press, Baltimore, Maryland, 1-13

Industries Assistance Commission (1978) Rural income fluctuations. Report no. 161. Aust Govt Pub Svc, Canberra

Industries Assistance Commission (1985) Crop and rainfall insurance. IAC circular 57/85

Industries Assistance Commission (1986) Crop and rainfall insurance. Report no. 393. Aust Govt Pub Svc, Canberra

Innes R (1988) Ex ante informational asymmetries and financial contracting. Univ California at Davis, Davis, California, mimeo

Innes R (1990) Limited liability and incentive contracting with ex ante action choices. J Econ Theory 52.1:45-57

Jodha NS (1981) Role of credit in farmers' adjustment against risk in arid and semiarid tropical areas of India. Econ Pol Weekly 16:1696-1709

Kotowitz Y (1987) Moral hazard. In: Eatwell J (ed) The new palgrave: a dictionary of economics. Macmillan Press, London, 549-551

Kramer RA (1983) Federal crop insurance 1938-1982. Agr Hist 57.2:181-200

Kunreuther H, Sanderson W, Vetschera R (1985) A behavioral model of the adoption of protective activities. J Econ Behav and Organization 6.1:1-15

Langemeier MR, Patrick GF (1990) Farmers' marginal propensity to consume: an application to Illinois grain farmers. Amer J Agr Econ 72.2:309-325

Lloyd AG, Mauldon RG (1986) Agricultural instability and alternative government policies: the Australian experience. In: Hazell P, Pomareda C, Valdes A (eds) Crop insurance for agricultural development, issues and experience. Johns Hopkins Univ Press, Baltimore, Maryland, 156-177

Lopes MR, Dias GL (1986) The Brazilian experience with crop insurance programs. In: Hazell P, Pomareda C, Valdes A (eds) Crop insurance for agricultural development, issues and experience. Johns Hopkins Univ Press, Baltimore, Maryland, 240-262

McCloskey DN (1976) English open fields as behavior towards risk. In: Unselding P (ed) Research in Economic History. JAI Press, Greenwich, 124-170

Muyco RD (1987) Philippines. Crop Insurance in Asia. Asian Productivity Organization, Tokyo, 199-220

Myers RJ (1988) The value of ideal contingency markets in agriculture. Amer J Agr Econ 70.2:255-267

Nelson C, Loehman E (1984) Further toward a theory of agricultural insurance. Staff paper 84-17. Purdue University, West Layafette, Indiana, 26

1990 Farm Bill, Proposal of the Administration (1990), forward by Clayton Yeutter. Office of Publishing and Visual Communications, Washington, DC, 147

Patrick GF (1988) Mallee wheat farmers' demand for crop and rainfall insurance. Aust J Agr Econ 32.1:37-49

Patrick GF, Lloyd AG, Cary JW (1985) Crop and rainfall insurance: some results from the Victorian mallee. Staff paper, 85-14, 44. Purdue University, West Lafayette, Indiana

Paxson CH (1988) Household savings in Thailand: responses to income shocks. Discussion paper #137, Woodrow Wilson School of Public and International Affairs

Peck AE (1977) Implications of private storage of grains for buffer stock schemes to stabilize prices. Food Research Institute Studies 16:125-140.

Pomareda C (1986) An evaluation of the impact of credit insurance on bank performance in Panama. In: Hazell P, Pomareda C, Valdes A (eds) Crop insurance for agricultural development, issues and experience. Johns Hopkins Univ Press, Baltimore, Maryland, 101-114

Pope JP (1937) Crop insurance in U.S. Speech of United States Senator from Idaho. Washington Star, Washington, DC, March 1, 1937.

Roberts RAJ, Gudger WM, Gilboa D (1989) Crop insurance. FAO Agr Services Bulletin 78. Food and Agriculture Organization of the United Nations, Rome, 51

Rosenzweig MR (1988) Risk, implicit contracts and the family in rural areas of low-income countries. Econ J 98:1148-1170.

Rothschild M, Stiglitz J (1976) Equilibrium in competitive insurance markets: an essay on the economics of imperfect information. Quart J Econ 90:629-49

Roumasset JA (1978) The case against crop insurance in developing countries. The Philippine Review of Business and Economics, March issue, 87-107

Schacht H (1990) Sunflower snafu could hurt crop insurance program. San Francisco Chronicle (May 15, 1990):cb

Sen AC (1987) India. Crop insurance in Asia. Asian Productivity Organization, Tokyo, 159-166

Skees JR, Reed MR (1986) Crop insurance and adverse selection. Amer J Agr Econ 68.3:653-659

Technical Board for Agricultural Credit, Manila (1987) A case report on mutual aid. Crop insurance in Asia. Asian Productivity Organization, Tokyo, 221-228

U.S. Congress, Commission for the Improvement of the Federal Crop Insurance Program (1989) pp. 57-58

U.S. Congress, Commission for the Improvement of the Federal Crop Insurance Program (1989) Recommendations and findings to improve the Federal crop insurance program. Principal report, Washington, DC, 240

U.S. Department of Agriculture, Federal Crop Insurance Corporation (1981) Annual report to Congress—1980, Federal Crop Insurance Corporation. USDA, Washington, DC, 32

U.S. General Accounting Office (1988) Crop insurance, participation in and costs associated with the Federal program. Briefing report to congressional requesters. U.S. General Accounting Office/RCED-88-171BR. G.A.O., Washington, DC, 51

U.S. General Accounting Office (1987) Crop insurance, overpayment of claims by private companies costs the government millions. Briefing report to congressional requesters. U.S. General Accounting Office/RCED-88-7. G.A.O., Washington, DC, 72

U.S. General Accounting Office (1988) Crop insurance, program has merit, but FCIC should study ways to increase participation. Report to the Chairman, Committee on Agriculture, House of Representatives. U.S. General Accounting Office/RCED-88-211BR. G.A.O., Washington, DC, 10

U.S. General Accounting Office (1989) Disaster assistance, crop insurance can provide assistance more effectively than other programs. Report to the Chairman, Committee on Agriculture, House of Representatives. U.S. General Accounting Office/RCED-89-211. G.A.O., Washington, DC, 37

Walker TS (1980) Decision making by farmers and by the National Agricultural Research Program on the adoption and development of maize varieties in El Salvador. Ph.D. dissertation, Stanford University

Walker TS, Jodha NS (1986) How small farm households adapt to risk. In: Hazell P, Pomareda C, Valdes A (eds) Crop insurance for agricultural development, issues and experience. Johns Hopkins Univ Press, Baltimore, Maryland, 17-34

Williams JC, Wright BD (1991) Storage and commodity markets. Cambridge Univ Press, New York, New York

Wright BD, Williams JC (1982) The roles of public and private storage in managing oil import disruptions. Bell J Econ 13:341-353

Yamauchi T (1986) Evolution of the crop insurance program in Japan. In: Hazell P, Pomareda C, Valdes A (eds) Crop insurance for agricultural development, issues and experience. Johns Hopkins Univ Press, Baltimore, Maryland, 223-239

Part II

Conceptual Issues

Chapter 5

The Optimal Design of Crop Insurance

J. QUIGGIN[1]

Risk and uncertainty are ever-present in agriculture. Government intervention at least ostensibly aimed at risk-reduction is almost equally ubiquitous. Action with respect to price uncertainty is most prevalent. Governments around the world have adopted a wide variety of policies aimed at stabilizing prices received by farmers for their products. For many farmers, however, yield uncertainty arising from climatic fluctuations is at least as important a source of risk as price uncertainty. Public policies aimed at compensating for yield uncertainty have been less widespread, and, at least at first sight, less successful, than policies aimed at stabilizing prices.

The main policy instrument designed to deal with yield uncertainty has been multiple peril crop insurance. In the United States, such a system has been in operation since 1939, though on a very limited basis until fairly recently. Canada has also operated a multiple peril crop insurance scheme. Other countries in which such schemes have been introduced include South Africa and a number of Latin American countries, while the issue has recently been the subject of active debate in Australia.

In nearly all cases, however, experience with multiple peril crop insurance has been disappointing. In the United States, indemnity payments under the scheme have consistently exceeded premium income, even in years of good weather conditions. In no country have premiums been consistently sufficient to cover both premiums and administrative costs.

Considerable analysis has been devoted to the properties of multiple peril crop insurance, and to proposals to improve the efficiency with which the existing scheme operates. Recent

---

[1]Research School of Social Sciences, Australian National Univerity, Canberra, Australia.

theoretical studies have included those of Ahsan, Ali and Kurian (1982), Nelson and Loehman (1987) and Chambers (1989).

There have also been a number of proposals to replace the existing scheme with one based on variables exogenous to the individual farm, such as regional yields or observed climatic experience. The Australian Industries Assistance Commission proposed a regional yield insurance scheme in its Report on Rural Income Fluctuations. Proposals for rainfall insurance were debated by Bardsley (1986) and Quiggin (1986).

In both of these classes of insurance, available information is not taken into account in the specification of the insurance contract. Multiple peril crop insurance schemes take no account of public information about relevant exogenous variables such as rainfall levels and regional yields. Conversely, schemes based on exogenous variables take no account of observed farm level yield outcomes. In this paper, a model is presented which is capable of encompassing both multiple peril crop insurance and insurance based on exogenous variables. The object is to derive a characterization of the optimal scheme in the presence of problems of moral hazard and adverse selection.

5.1 Problems of Crop Insurance

Multiple peril crop insurance has several features which tend to make the problems of insurance particularly acute. These difficulties may be reinforced by consideration of the only type of yield risk for which commercial insurance is readily available, namely hail damage.

5.1.1 Absence of pooling. The first way in which the crop insurance problem differs from other insurance problems is that there is little potential gain from the pooling of similar risks. Whereas in cases such as life and motor vehicle insurance, individual risks are uncorrelated, in the crop insurance case, all the farms in a given district are likely to experience adverse conditions at the same time, and correlations are likely to be high even for large regions. Among other things, this is an explanation of why multiple peril crop insurance has not emerged spontaneously from arrangements among groups of risk-bearing individuals in the way in which other forms of insurance have done.

The absence of risk pooling is not a fatal objection, however. It leaves open the possibility that organizations or governments with a large portfolio, having returns uncorrelated with production risk, will offer insurance. Provided the amount of insurance liability is small in relation to the total portfolio this can be done without charging a risk premium.

If large-scale multiple peril crop insurance is to be offered, however, only governments are likely to have a sufficiently large portfolio. This creates a number of theoretical and practical problems. First, if agriculture is sufficiently important in the economy, government revenue will be positively correlated with yields. Second, the involvement of government on the efficiency grounds suggested here opens the way to various inefficient forms of subsidization.

5.1.2 Moral Hazard. The moral hazard problem in multiple peril crop insurance arises from the fact that farmers can take a great many actions which affect their final yield. For example, they can choose whether or not to apply pesticides and, if so, how much. They can choose varieties which have high yields but are highly susceptible to drought or insect attack, or, instead, more robust but lower-yielding varieties. More subtle factors such as the care with which soil preparation and ploughing is undertaken may also have an impact.

In all of these cases, a farmer who has taken out multiple peril crop insurance will face lower losses in the event of crop failure. This means that it will be rational to choose techniques which involve a greater risk of failure. Consider the choice between high yielding and robust varieties. An insured farmer will tend to choose the high yielding variety. If the season is good, the farmer reaps the benefit. If it is bad, the insurer bears part of the cost.

As in the simple example above, the problem of moral hazard means that the insurer must charge higher rates, reducing the viability of insurance for those farmers who do not plan to adopt risky techniques. There are a variety of ways in which the insurance company may seek to alleviate the problem of moral hazard. Contracts may specify that farmers should use appropriate techniques and should not take unreasonable risks. Unfortunately, the difficulty of specifying these techniques is greater than in standard moral hazard problems, such as burglary and fire insurance. There are few simple techniques for reducing the risk of crop failure which can be specified in a contract. For example, an attempt by insurance companies to dictate farmers' choice of variety would probably not be considered acceptable by many farmers and would require continuous revision of contracts as new and modified varieties became available.

Even where such requirements could be specified, enforcing them would be very costly. Suppose, for example, that certain applications of pesticides were specified in the contract. Unless the insurance company had agents stationed on farms, it would be very difficult to check that the applications were being carried out. Any attempt to refuse payment on the grounds that farmers had violated such contract provisions would certainly end in expensive litigation.

5.1.3 Adverse Selection. Several types of adverse selection problems arise in connection with multiple peril crop insurance. The typical contract permits farmers to choose, on an annual basis, whether to insure and if so, which crops they should insure. From the simple example given above, it is apparent that adverse selection problems may arise if farmers (or farms) differ in ways which are not reflected in the premium structure.

Consider first the case where the insurance contract specifies a payout whenever yields fall below a certain proportion of the historical average for the county. Then farmers whose yields are generally lower than average (either because of poor soil or poor management) will have an incentive to insure, while farmers whose yields are better than average will have an incentive not to insure.

There has been a partial response to this problem by permitting those farmers with records of yields to insure on the basis of their own farm yields rather than the county average. This may help to induce better than average farmers to insure, but it is obvious that worse than average farmers will prefer to insure on the basis of the county average.

The fact that farmers may choose annually whether or not to renew their insurance presents a second possibility of adverse selection. If the decision is left open sufficiently late, it may be possible for farmers to predict whether a drought is likely. Given the current reliability of long-range weather forecasting this is unlikely to be a problem unless the closing date for insurance decisions is very late indeed.

A third form of adverse selection arises from the fact that farmers can choose to insure some crops but not others. Obviously this also means that, by making appropriate planting decisions, they can choose to insure some fields but not others. A farmer with land of varying quality might choose to plant wheat on the best land and maize on the worst (that is, the land most susceptible to loss) then insure the maize but not the wheat.

5.1.4 Imperfect Indemnity. Imperfect indemnity from multiple peril crop insurance arises in two basic forms. First, the insurance payment may fail to cover the losses incurred by the farmer (measured by the differences between actual and expected yield at the price expected at the time the contract is sold). This is likely to occur if the contract is modified so as to reduce the incidence of moral hazard and adverse selection, for example, by the inclusion of a deductible. The problem of imperfect indemnity becomes even more severe when proposals such as rainfall insurance, which take no account of observed farm-level yields, are considered.

The second and more fundamental problem is that even if the insurance payout exactly covers losses due to crop failure these will not be perfectly correlated with variations in farmers' incomes. Other major sources of income variation include variations in output prices, costs, yields for uninsured commodities (most notably livestock) and variations in off-farm income.

Variations in output prices are the most important source of concern. The relative contribution of price and yield variations to income instability has been examined by a number of authors including Piggott (1978). For some commodities and regions (for example, Australian wool), price variations are dominant and for other commodities (for example, Australian wheat) yield variations are dominant.

In the case of North American cropping industries, it is also necessary to take account of the fact that yields and prices are negatively correlated. In years of low yields, the excess of demand over supply tends to drive prices up. Depending on the relevant elasticities, this may mean for some farmers that yields are negatively correlated with net farm incomes.

5.1.5 Hail Insurance—A Special Case. The great exception to this generally negative picture has been hail insurance. Privately run hail insurance schemes have operated successfully in a number of countries. The conclusion might be drawn that, given efficient business management, a similarly favorable outcome could be expected for multiple peril crop insurance.

Unfortunately, there are a number of factors specific to hail damage which make it much more favorable candidate for insurance than yield risks in general. The most obvious, but perhaps the least important, special feature is that hail damage risks are amenable to pooling. Given a moderate spread of locations the likelihood that a large proportion will suffer hail damage in any one year is fairly small. This fact is important in explaining the spontaneous emergence of hail insurance. It is not, however, a major objection to the feasibility of a government-backed insurance scheme for risks which cannot be pooled across farmers, since the portfolio of risks held by governments is large and diverse.

The second feature is the absence of major moral hazard problems. There is nothing that can be done to prevent a hailstorm occurring and comparatively little that can be done to mitigate its impact. This makes hail insurance more attractive than insurance against other specific risks such as insect damage, where moral hazard problems may be serious. Multiple peril crop insurance has even greater problems, since no specific cause of crop losses need be demonstrated.

Adverse selection problems are also unlikely to be serious. While some localities are more hail-prone than others, and some crops are more susceptible to damage than others, these facts can easily be taken into account in setting the rates. Finally, the absence of a correlation between hail damage for farmers in different regions means that prices are unlikely to be negatively correlated with hail damage. This reduces to some extent the problems of imperfect indemnity referred to above.

Somewhat similar factors have led to the successful development of private insurance markets against excess rainfall for a number of agricultural industries (for example, tomatoes) and also for some non-agricultural industries (outdoor entertainments). It is not immediately clear why insurance is privately available against excessive rainfall, but not against deficient rainfall. This question is examined in the following section.

5.2 Alternatives to Multiple Peril Crop Insurance

The problems of moral hazard and adverse selection make it unlikely that a multiple peril crop insurance scheme can be operated without substantial and continuing subsidies. The Australian debate over policy measures to cope with rural income fluctuations has raised a number of suggestions which may be worth considering. In this section attention will be focused on rainfall insurance.

5.2.1 Rainfall Insurance. The basic idea of rainfall insurance is that insured farmers should receive a payout whenever rainfall in their area falls below a specified level. The insurance payout would depend only on rainfall and the farmer would not be required to demonstrate any loss. Indeed, there is no reason in principle to require that persons taking out insurance should be farmers at all.

Since there is nothing which can be done to influence the probability of rainfall, the problem of moral hazard is eliminated, just as it is for hail insurance. Similarly, given the current state of long-term weather forecasting, there are no problems of adverse selection. (Even if droughts were predictable in advance, this could be dealt with by allowing premiums to vary from year to year.) Finally, administration costs are likely to be low.

In these respects, rainfall insurance is similar to hail insurance, which has already proven itself to be commercially viable. There are, however, some difficulties which arise in the case of rainfall insurance and these were the subject of debate between Bardsley, Abey and Davenport (1984) and Quiggin (1986).

The first problem is that, unlike hail, drought experience is highly correlated across farmers. Bardsley, Abey and Davenport suggested that this would significantly reduce the viability of rainfall insurance. However, Quiggin argued that their model dealt with risks to government in an inappropriate fashion, and that at least while the scheme was moderate in scale, government could be regarded as risk-neutral.

The second problem is that of imperfect indemnity. Because payouts are based on rainfall rather than on observed yields, the indemnity properties of rainfall insurance are worse than those of multiple peril crop insurance. Bardsley, Abey and Davenport argued that this would make the scheme less attractive than self-insurance, at least in the presence of moderate administrative costs. However, their model specification was criticized on this point by Quiggin.

While the debate did not reach a settled conclusion, there was a consensus that a rainfall insurance scheme would not have a major impact in the absence of some subsidy at least on administrative costs. On the other hand, if subsidies were to be paid to farmers suffering adverse climatic conditions, rainfall insurance would be one of the most cost-effective alternatives.

A number of modifications of rainfall insurance could be proposed to improve its indemnity properties. Crop growth models incorporating rainfall and temperature data could be used to determine the payout on the basis of scientific estimates of likely best practice yields, rather than on arbitrary rainfall levels. Also, price fluctuations could be taken into account along the lines suggested in the Industries Assistance Commission (1977) regional yield insurance scheme.

An issue which was not considered in the course of the debate is the interaction between crop insurance and futures markets. Since the possibility of a yield shortfall is a major deterrent to forward selling, these two options tend to complement each other.

Neither a multiple crop insurance scheme nor a rainfall insurance scheme will in general be optimal. Existing multiple peril crop insurance schemes do not take account of public information such as rainfall levels. This exacerbates the well-known problems of moral hazard and adverse selection. Conversely, proposed rainfall insurance schemes do not take account of the information contained in observed yields and therefore do not maximize the amount of risk-reduction which can be offered.

It is, therefore, desirable to consider the optimal design crop insurance. The problem may be analyzed in the framework of agency theory. The principal in this case is the insurer (assumed to be risk-neutral), while the agent is the insured farmer. A formal approach to the problem is presented in the following section.

5.3 The Model

Choices under uncertainty may be described in terms of an act-state model. Attention will be confined to the case of a finite set of states S, although it is straightforward to extend the analysis to the case of a general σ-field of events. Each act A has consequences which may vary depending on the state of the world which occurs. Decisions are said to be made under uncertainty if the state of the world and, hence, the consequences are not known at the time the decision is made.

The standard theory of choice under uncertainty, Expected Utility theory, is based on the assumption that the decision-maker knows the probability of each of the possible states of the world and also the consequences of each possible act in each possible state of the world. Most of the theory deals with the case where consequences can be expressed in terms of income or wealth levels, denoted x. Further, to each possible outcome x, the decision-maker attaches a number U(x) representing the utility of that outcome. Given an act A which has consequence x_i in state S_i the expected utility from that act is

$$EU(A) = \sum_{i=1}^{n} p_i \, U(x_i), \tag{1}$$

where p_i is the probability of state S_i. Given a choice between a number of acts the decision-maker will choose the one which maximizes EU(A).[2]

The crop insurance problem may be modelled as follows. The insurer can observe 'public' information about the state of the world, such as rainfall levels and county average yields. The public information on the state of the world is given by a variable θ, which may take *N* possible values. The full state of the world

---

[2]The assumption implicit in the functional form (1) that preferences are 'linear in the probabilities' is excessively restrictive and can easily be modified (see Fishburn 1988 for a number of possible generalizations). Quiggin and Fisher (1989) discuss the implications for problems of stabilization and insurance.

is given by θ and an additional random variable ψ, which cannot be observed. The bivariate probability distribution of θ and ψ is given by $F(\theta, \psi)$. In general θ and ψ are not independent and for each value of θ, there is a conditional probability distribution $F_\theta(\psi)$.

Farmers are assumed to grow a single crop, and to obtain a yield X. Yield is a random variable depending on the state of nature (θ, ψ) and also on other variables such as the farmer's effort, ability and input choices. These variables are known to the farmer, but not to the insurer. The insurer may observe X, although this may involve some administrative cost. The public information available to the insurer is thus given by X and θ.

The insurer and the insured must agree on a contract (disagreement will result in a zero contract). It will be assumed that because of competition, the insurer earns zero profits, and the insured chooses the most preferred contract consistent with this condition. The general contract specifies a premium π, and a payment $\rho(\theta, X)$ dependent on the public information about the state of the world and the farmer's observed yields.

Both adverse selection and moral hazard problems will arise in general. The problem of moral hazard arises because the insurer cannot observe the farmer's effort level. The problem of adverse selection will arise because the insurer has less information than the farmer concerning the riskiness of the farm's operations and the expected yield in the absence of insurance. The case of moral hazard will be considered first.

5.3.1 Moral Hazard. The moral hazard problem may be modelled as follows. The yield level X is determined by θ, ψ and the farmer's effort level e.

The farmer's income is given by

$$Y = rX - C(e) - \pi + \rho(\theta,X) + \varepsilon, \tag{2}$$

where r is the output price (here assumed to be non-stochastic) and ε is off-farm income and income from non-insured activity. Given a particular contract, the farmer's decision variable is the effort level, e, and the objective is to maximize E[U(Y)]. The first order condition is given by

$$E[U'(Y)\{(r + \partial\rho/\partial X)\partial X/\partial e - C'(e)\}] = 0. \tag{3}$$

The problem of designing an optimal contract may be approached by splitting the problem into two parts. The first is to design the optimal contract contingent on a given observed value of θ. Such a contract may be specified by a premium π_θ and a contingent payment schedule $\rho_\theta(X)$. Since all information apart from the value of θ is private to the farmer, this problem is equivalent to that studied in the standard moral hazard model. The independence axiom of Expected Utility theory implies that the contingent contract selected in advance for each value of θ will be the same as that which would be selected if the state of the world were known to be θ at the time the contract was made.[3] The second part of the problem is to design the complete contract, consisting of a premium π and a set of state-contingent payments to the farmer $\rho(\theta)$, along with the state-contingent insurance contracts derived in the first stage.

Consider now the derivation of the contract contingent on a given value of θ. Yield is given by a function $X_\theta(\psi, e)$ depending on the unobservable state of the world and the farmer's effort. If state θ is realized, the farmer will receive a payment $\rho(\theta)$, so that income in the absence of a contingent insurance contract is given by

$$Y_\theta = rX_\theta(\psi,e) - C(e) + \rho(\theta) - \pi + \varepsilon. \tag{4}$$

---

[3]This result will not apply in more general models of choice under uncertainty such as those of Machina (1982) and Quiggin (1982) in which the Independence Axiom does not apply.

We will consider the case where the contingent contract set consists of a range of deductibles (K). The agent pays a premium π_θ and receives a payment

$$\rho_\theta(X, K) = \max\{(ER - R) - K, 0\} \tag{5}$$

where $R = rX_\theta(\psi, e)$.

The optimal level of the deduction (assuming that any level of contingent insurance is selected) is determined by

$$\max_K E[U(Y_\theta - \pi_\theta + \rho_\theta(X,K))], \tag{6}$$

subject to conditions

$$E[\pi_\theta - \rho_\theta(X,K)] = c + s, \tag{A}$$

where c represents administrative costs such as those of observing the insured farmer's output, and s is the net subsidy paid by the government, and

$$E[U'(Y_\theta)\{(r + \partial\rho/\partial X)\,\partial X/\partial e - C'(e)\}] = 0. \tag{B}$$

Condition (A) is a zero profit condition implied by risk neutrality. Condition (B) ensures that the farmer chooses the most preferred level of effort given the selected insurance contract.

If the administrative costs are high or the problem of moral hazard is severe, the alternative optimum will be selected in which there is no contingent insurance and the farmer receives the income Y given by (4).

We may now consider the general problem. This is to choose the non-contingent premium π and the set of non-contingent payments $\rho(\theta)$, with the associated contingent contracts given by (5), so as to maximize E[U(Y)] subject to the condition

$$E[\rho(\theta)] = \pi. \tag{7}$$

It was observed above that the optimal contingent contract, given a particular value of θ, is the same as that which would be chosen if the value of θ were known, and is independent of the insurance contract as a whole. It is not generally true, however, that the optimal set of payments $\rho(\theta)$ is independent of the available set of contingent contracts. To see this consider the case when θ can take only two values, 0 and 1. Assume first that X is independent of the unobservable effort variable so that full contingent insurance against variations in Y is possible, at actuarially fair rates, in both states. It is clear that the payments $\rho(\theta)$ will be chosen so as to equalize income across the states. Now suppose that when $\theta = 0$, full contingent insurance is possible, but that when $\theta = 1$, moral hazard prevents the emergence of an insurance market. If $\rho(\theta)$ were chosen so as to equalize expected income across the states, then marginal utility of income in state 0 is simply $U'(Y_0)$. Expected marginal utility of income in state 1 is $E[U'(Y_1)]$, where $E[Y_1] = Y_0$. If U' is linear in Y, that is, if the utility function is quadratic, then $E[U'(Y_1)] = U'(Y_0)$. In the more plausible case where $U''' > 0$, and U' is a convex function, $E[U'(Y_1)] > U'(Y_0)$ by Jensen's inequality. Hence, the optimal solution will have higher expected marginal utility of income in the state where insurance is unavailable. More generally, if the contingent distribution, after taking into account the optimal contingent insurance contract, is riskier (in the sense of Rothschild and Stiglitz 1970, 1971 and 1974) in state 1 than in state 0, the optimal solution will have higher expected marginal utility of income in state 1.

The analysis presented above may now be related to the choice between multiple peril crop insurance and rainfall insurance. Suppose that the public information θ is given by rainfall levels, and that all other relevant aspects of uncertainty are captured by ψ. First consider the case where moral hazard precludes contingent insurance for all values of θ. In this case, the optimal contract is a pure rainfall insurance contract. Conversely, a standard multiple peril crop insurance contract will be optimal when the optimal deductible is independent of θ. An intermediate option which could also arise is one in which a payout equal to some proportion of losses is made, subject to the requirement that θ lie within a particular set of values. For example, farmers could be compensated for their actual losses whenever rainfall fell below some trigger level. The standard defined-risk insurance contracts such as hail insurance fall into this category.

5.3.2 Adverse selection. The case of adverse selection will be considered next. The variable unobservable by the insurer is not an effort level chosen by the insured farmer, but exogenously determined farmer quality or land quality. For convenience, it may be assumed that there are two classes of farmers—high risk and low risk. In the standard analysis, without contingent contracts, there are three possible cases—pooling equilibria in which both groups are insured on the same terms, separating equilibria in which two contracts are offered and the two groups self-select the contracts designed for them, and cases in which there is no equilibrium involving insurance. The separating equilibrium has the property that the high-risk group are fully insured at actuarially fair rates, while the low-risk group are only partially insured.

The possibility of contingent contracts yields different results. Consider the case when there are two possible values of the observable variable θ. When $\theta = 0$, low-risk farmers face zero risk of loss, while high risk farmers have a positive risk. When $\theta = 1$ both groups risk loss, with a higher probability for the high-risk group. If the value of θ were known to be 0, the high risk group would be the only one to seek insurance and, given competition among insurers, would receive it at actuarially fair rates. If the value of θ were known in advance to be 1, then any of the possibilities described by the standard analysis could arise.

In the contingent contract case, people who take out insurance contracts contingent on the state $\theta = 0$ identify themselves as members of the high-risk group. Thus, an insurer who offers insurance contingent on the occurrence of state $\theta = 1$ can effectively discriminate against the high-risk group. If the probability of state $\theta = 0$ is sufficiently high and the high-risk group is sufficiently risk-averse, the insurer can offer the low-risk group actuarially fair insurance contingent on state $\theta = 1$. Even though the contingent contract is actuarially favorable to them, the high-risk group will prefer to buy insurance which is not contingent on state $\theta = 1$.

This equilibrium will be stable only if the insurer can prevent high-risk individuals from purchasing the contingent contract and separately insuring against losses in the state $\theta = 0$. In this context, it may be observed that most insurance contracts contain clauses requiring notification of any other insurance contract covering the insured risk.

The adverse selection problem may be modelled as follows for the case of two risk classes and N values of the public state variable θ. The farmer's yield X is determined by θ, ψ and the risk class i, which may take the values 1 or 2. The high-risk class is denoted 2. In order to maintain comparability with the standard analysis (see, for example, Laffont 1989), it is assumed that X may take two values: X_L corresponding to loss, and X_H corresponding to no loss. Denote the probability of loss for agents of type i by

$$p^i = \Pr\{X(\theta, \psi, i) = X_L\}. \tag{8}$$

The farmer's output contingent on the state θ is denoted by $X_\theta(\psi, i)$ as before. The probability of loss for farmers of type i contingent on the public state θ is similarly denoted by

$$p^i_\theta = \{\Pr\{X_\theta(\psi, i) = X_L\}. \tag{9}$$

We may now analyze the conditions for existence of a separating equilibrium. In order for such an equilibrium to exist, the high-risk group must be fully insured at actuarially fair rates. The problem is to specify the most attractive contract for the low-risk groups, subject to the requirements that it be actuarially fair for the high risk and that it should not be preferred by the high-risk group to the contract designed for them. Since there are only two possible outcomes for the yield variable, a contract may be specified by stating the premium π and the payment $\rho(\theta)$ which is made contingent on state θ, subject to the occurrence of the loss value X_L. Thus for the high risk group we have $\rho^2(\theta) = r(X_H - X_L)\ \forall\ \theta$, an actuarially fair full insurance contract.

For the low risk group, the problem may be stated as follows. Let

$$Y^i_\theta = rX_\theta(\psi, e) - \pi^i + p^i(\theta) \qquad i = 1, 2 \tag{10}$$

be income after insurance. Then the problem is to choose the vector $\rho(\theta)$ to maximize $E[U(Y^1)]$ such that

(A) $\quad E[\rho^1(\theta)] = \pi^1$

(B) $\quad E[U(Y^2_\theta)] \geq E[U(rX^2_\theta - \pi^1 + \rho^1)]$

(C) $\quad E[U(Y^1_\theta)] \geq E[U(rX^1_\theta - \pi^2 + \rho^2)]$

(D) $\quad 0 \leq \rho^1(\theta) \leq r(X_H - X_L)\ \forall\ \theta$

The first inequality constraint, specifying that the high-risk contract be at least as attractive to the high-risk group as the low-risk contract, will, in general, be binding. Competition to make the low-risk contract attractive to the low-risk group will proceed up to the point where high-risk purchasers are about to enter the market. The second inequality will not normally be binding if a separating equilibrium emerges. The final set of inequalities may or may not be binding. Attention will initially be confined to the case where these constraints are non-binding. The first-order conditions are:

$$p^1_\theta(U'(Y^1_\theta + \lambda_1) - \lambda_2 p^2_\theta\, U'(Y^2_\theta) = 0 \tag{11}$$

$$U'(Y^1_\theta) = \lambda_2 p^2_\theta \,/\, p^1_\theta\, U'(Y^2_\theta) - \lambda_1 \tag{12}$$

Because the high-risk group is fully insured at actuarially fair rates, $U'(Y_\theta^2)$ is a constant independent of θ. Thus, $U'(Y_\theta^1)$ depends linearly on p_θ^2 / p_θ^1, the relative probability of a loss for the high-risk group. The higher is p_θ^2 / p_θ^1, the higher is $U'(Y_\theta^1)$ and hence, by concavity, the lower is Y_θ^1. Thus, as in the standard analysis, the greater the difference between the high and low-risk groups, the larger the deductible the low-risk group must accept in the separating equilibrium. If p_θ^2 / p_θ^1 is sufficiently large, the constraint $0 \leq \rho^1(\theta)$ becomes binding, and the low-risk group receives no insurance in this state.

However, the interpretation of this case in the contingent model is quite different from that of the standard analysis. The existence of states in which the low-risk contract offers no insurance serves to make that contract less attractive to the high-risk group, and thereby relax the constraint (B), implying the possibility of more complete coverage for the low-risk group in other states.

In the crop insurance context, the crucial issue is to identify the likely variation of p_θ^2 / p_θ^1 across states. This variable measures the relative vulnerability of good and poor managers to the various events represented by the public state variable θ. For some events, most notably hailstorms, it seems reasonable to postulate a value of p_θ^2 / p_θ^1 very close to 1. For events of this kind, a pooling equilibrium will naturally emerge. Further, since insurance choices regarding such risks convey little information about a farmer's management ability, there is no incentive to bundle hail insurance into a more general insurance contract. This is in accordance with the observed market outcome. Insect attack is an example of an event for which vulnerability to loss is likely to be highly dependent on management availability. Drought is an intermediate case. There is scope for management to alleviate the consequences of moderate shortfalls in rain, but there is little that can be done in the case of a severe drought. The argument above indicates that the availability of actuarially fair multiple peril crop insurance (the option likely to be preferred by the high-risk group) would assist in the achievement of a separating equilibrium in a market for insurance contingent on the occurrence of a drought.

5.4 Concluding Comments

The existence of observable information affecting the likelihood of crop losses poses problems for the design of insurance mechanisms which have not been considered adequately in the policy debate. The present chapter has outlined some of the characteristics of an optimal scheme in situations of moral hazard and adverse selection.

The representation of the optimal scheme in terms of contingent insurance contracts yields some new and interesting results. In the moral hazard case, the optimal payment on the observation of different values of the public state variable depends on the properties of the contingent insurance contract against private risk. In the adverse selection case, the existence of different public states provides new opportunities for the development of separating equilibriums.

References

Ahsan SM, Ali AG, Kurian NJ (1982) Toward a theory of agricultural insurance. Amer J Agric Econ 64,3:520-529

Bardsley P (1986) A note on the viability of rainfall insurance—reply. Aust J Agric Econ 30:70-72

Bardsley P, Abey A, Davenport S (1984) The economics of insuring crops against drought. Aust J Agric Econ 28:1-14

Chambers R (1989) Insurability and moral hazard in agricultural insurance markets. Amer J Agric Econ 71:604-616

Fishburn P (1988) Nonlinear preference and utility theory. Johns Hopkins Univ Press, Baltimore, Maryland

Industries Assistance Commission (IAC) (1978) Report on rural income fluctuations. AGPS, Canberra

IAC (1977) See p. 10

Laffont JJ (1980) Essay in economics of uncertainty. Howard Univ Press, Cambridge, Massachusetts

Laffont JJ (1989) The economics of uncertainty and information. MIT Press, Cambridge, Massachusetts

Machina M (1982) Expected utility analysis without the independence axiom. Econometrica 50:277-323

Nelson C, Loehman E (1987) Further toward a theory of agricultural insurance. Amer J Agric Econ 69:524-531

Piggott R (1978) Decomposing the variance of gross revenue into demand and supply components. Aust J Agric Econ 22:145-156

Quiggin J (1982) A theory of anticipated utility. J Econ Behav and Organization 3:323-343

Quiggin J (1986) A note on the viability of rainfall insurance. Aust J Agric Econ 30:63-69

Quiggin J, Fisher BS (1989) Generalized utility theories: implications for stabilization policy. Aust Agric Econ Soc conference, Christchurch

Rothschild M, Stiglitz J (1974) Equilibrium in competitive insurance markets: an essay in the economics of imperfect information. Quart J Econ 90:629-649

Rothschild M, Stiglitz J (1970) Increasing risk: I. A definition. J Econ Theory 2:225-243

Rothschild M, Stiglitz J (1971) Increasing risk: II. Its economic consequences. J Econ Theory 3:66-84

Chapter 6

Agricultural Insurance, Production and the Environment

R. INNES[1] and S. ARDILA[2]

6.1 Introduction

Using a simple model of production under uncertainty, this paper studies the effects of agricultural insurance on farm production choices, soil depletion and the environment. Unlike prior studies of production under uncertainty, this paper allows for two related risks, one in short-to-medium run production revenues and the other in asset/land value. The analysis characterizes effects of insurance programs which stabilize either, both or linear combinations of short-run revenue risk and land price risk. In addition, the paper illustrates the prospective role and importance of land price risk in the determination of insurance program impacts, while distinguishing between effects of "pure" stabilizing insurance and the "truncating" types of insurance that are observed in practice.

The model that we examine is a rather simple extension of standard two-date models of production under uncertainty (e.g., see Sandmo and the large literature that has sprung therefrom). In these models, the farmer chooses an ex-ante output level that yields a random ex-post revenue. The main innovations here are that: (i) ex-ante output depletes the soil stock, which leads to a lower end-of-period value of the farmer's land, and (ii) both production revenue and land price are risky. Due to land price risk, an investment in the soil stock, via a reduction in ex-ante output, is

---

[1]Department of Agricultural Economics, University of Arizona, Tucson, Arizona 85621, USA.

[2]Environmental Protection Division, Inter-American Development Bank, Washington, DC, USA.

risky. Moreover, if land price risk "dominates" production risk in a sense made precise below, then "investment" in the soil stock is more risky than "investment" in output; therefore, vis-a-vis optimal choices under farmer risk neutrality or no risk, risk and risk aversion will lead to less investment in the more risky soil stock "asset"—and hence, a higher output level. This conclusion is precisely the opposite of that derived in standard models of production under uncertainty, namely, that risk and risk aversion lead to lower optimal output levels. Inasmuch as higher farm outputs raise the environmental costs of agricultural production, this conclusion also indicates that the public may have a stake in reducing the risk faced by farmers beyond the stake that it may have in prospective risk-sharing benefits from agricultural insurance. The ensuing analysis studies whether and how the government may realize this stake by offering production-stabilizing and land-value-stabilizing insurance programs.

This study builds on research in three economics literatures: (1) production under uncertainty (e.g., see Saha, Innes and Pope (1991) and the references therein), (2) agricultural insurance (e.g., see Quiggin (1992), Chambers (1989), Miranda (1991), and the references therein), and (3) soil conservation (e.g., see LaFrance (1992), Ardila and Innes (1991) and the references therein). The first two literatures are extended by incorporating soil evolution effects of output choices and risk in the value of the terminal soil stock/land price. Note, however, that this paper abstracts from moral hazard and adverse selection problems that have been the object of study in recent research on agricultural insurance; this abstraction permits a focus on the primitive effects of insurance programs in a dynamic environment.[3]

---

[3]In defense of our no-information-problem assumption, note that information-based explanations for an absence of private insurance markets need not provide a motivation for government provision of insurance since government is subject to precisely the same moral hazard and adverse selection problems that confront potential private insurers. However, as John Quiggin (1992) points out, high cross-farm correlations in risk render private provision of insurance problematic, providing a potential motivation for governments, as the agents with the largest possible

In the soil conservation literature, Ardila and Innes (1991) study wealth and risk effects on production and soil depletion in two and three-date settings that, like this paper's model, incorporate both production and land value risk.[4] In the analysis that follows, we build on Ardila and Innes (1991) by characterizing effects of various stylized insurance programs on output, soil depletion and the environment.

The balance of the paper is organized as follows. The model is presented in Section II, followed by an analysis of "pure" insurance programs in Section III and traditional "truncating" insurance programs in Section IV. Finally, Section V discusses implications of incorporating explicit land value effects of insurance programs. Section VI gives some concluding remarks.

---

portfolios, to enter the breach. High cross-farm correlations in risk also imply that moral hazard and adverse selection problems can be overcome at rather little cost by conditioning insurance payments on regional outcomes, rather than farm-level outcomes (Arzac (1989), Miranda (1991)). Thus, while we would not want to diminish the importance of studying informational problems in insurance contract design, we believe that our neglect of these problems merely focuses attention on the effects of government insurance programs that have the most obvious imperfect market motivation, namely, those that insure correlated risks.

[4]In the soil conservation literature, production effects on soil evolution have been carefully modelled, but very little attention has been directed to the implications of uncertainty and farmer risk aversion for these effects. Excepting Ardila and Innes (1991), only two papers have (to our knowledge) considered soil conservation issues in uncertain environments, Kramer, McSweeney and Stavros (1983), and Stefanou and Shortle (1986). Both of these papers are concerned with the specification and simulation of a generalized mean-variance model which lacks behavioral underpinnings (see McSweeney and Kramer (1986)) and which is not used to derive the qualitative properties of optimal choices that are of interest here. General studies of renewable resource management under uncertainty (Burt (1964) and Pindyck (1984)) have also focused on different issues than does this paper; these studies examine implications of risk in stock regeneration (e.g., soil evolution) for optimal decision rules when farmers are risk neutral.

6.2 The Model

We consider a simple model of a farmer's production choices which incorporates soil stock effects and two different types of risk. The first type of risk is in production, as is standard in the extensive literature on production under uncertainty. The second type of risk is in land price or, equivalently, in the value of the end-of-period soil stock.

Formally, we assume that the farmer has an increasing and concave Von Neumann-Morgenstern utility function in end-of-period wealth W, V(W). We also assume that the farmer's aversion to risk declines with his wealth in the following sense:

$$\frac{d}{dW}\left(\frac{-V''(W)}{V'(W)}\right) < 0 \tag{1}$$

That is, V( ) exhibits Decreasing (Arrow-Pratt) Absolute Risk Aversion (DARA).

End-of-period wealth is the sum of (i) production profits, (ii) exogenous other income, and (iii) the sale value of the farmer's land, which depends on the end-of-period soil stock. For simplicity, we assume that the size of the farm (i.e., the land acreage owned by the farmer) is fixed. Production profits then depend on the farmer's choice of an ex-ante production target, Y, which yields (i) ex-post revenue of F(Y, ε), where ε is a non-negative random variable, and (ii) a cost (in units of end-of-period wealth) of $e(Y; X_1)$, where X_1 is the soil stock as of the initial time 1.[5] F( ) and e( ) are assumed to have the usual properties: $F_Y > 0$, $F_{YY} < 0$, $F_\varepsilon > 0$, $F_{Y\varepsilon} \geq 0$, $e_Y > 0$, and $e_{YY} > 0$. In addition, we

[5]Some of these production costs may be borne at time 1 and others at time 2. In deriving e( ), the production costs borne at time 1 are inflated by the interest rate to obtain an end-of-period equivalent.

will make the rather standard assumption that risk enters production in a linear fashion:[6]

$$F(Y, \varepsilon) = \alpha(Y) + f(Y)\varepsilon \tag{2}$$

Given Y and X_1, production profits are:

$$F(Y, \varepsilon) - e(Y; X_1) \tag{3}$$

The end-of-period land price depends on X_2, the end-of-period soil stock, which in turn depends on the production choice, Y, and the initial soil stock, X_1, according to the following equation of motion:

$$X_2 = X_1 + G - g(Y) \tag{4}$$

G is the autonomous growth in the stock; $g(\ ) > 0$ is the production-induced depletion in soil; and g is assumed to be increasing and weakly convex so that higher outputs are associated with higher levels of soil degradation. The third component of W, the land value, can now be defined as

$$\begin{aligned} L(X_2, \varepsilon_L, \theta) &= L(X_1 + G - g(Y_2), \varepsilon_L, \theta) \\ &= \beta_0(X_2) + \ell(X_2)\varepsilon_L + \beta_1\theta \end{aligned} \tag{5}$$

where $\varepsilon_L = \varepsilon$, θ is a non-negative random variable that is independent of ε and $L(\ )$ is assumed to be an increasing function, i.e., higher levels of the soil stock (and/or the random variables ε and θ) elicit a higher land price. In addition, we assume that (i) risk enters linearly in the land value relation, as indicated in (5), (ii) $\ell'(\ ) = L_{X\varepsilon} \geq 0$, so that higher values of ε do not imply lower

---

[6]Signing risk effects in models of uncertainty is notoriously difficult and invariably requires a linear risk specification. The specification in (2) is a multiplicative risk specification that is standard in models of production under uncertainty (e.g., see Just and Pope (1987)).

values of marginal soil stock, (iii) $L_{x\theta} = 0$, so that the random variable θ does not affect the marginal value of the soil stock, and (iv) $L_{xx} \leq 0$, so that the marginal value of soil is non-increasing. We subscript the ε_L argument in the land value function for analytical purposes that will be discussed momentarily.

Several features of this specification merit elaboration at this point. First, our soil stock can be interpreted as an index of soil productivity that is actually a function of many variables, including organic matter content, soil structure, PH, and contents of several minerals. For the sake of analytical simplicity, we posit a single index for these soil characteristics' effects. We also posit a positive relationship between land price and our soil-stock/soil-productivity-index variable based on both economic logic (i.e., more productive land yields higher revenue flows) and empirical evidence (e.g., see Miranowski and Hammes (1984), Gardner and Barrows (1985), and King and Sinden (1988)).

Second, production revenues and land values are clearly subject to some common shocks, including changes in output demand, input costs and technology. We model these common shocks with our random variable ε. Land values may also be sensitive to random events that are independent of those affecting production revenues; for example, land values may change with random demographic and population shifts. We capture this second set of random effects with our variable θ.[7]

Third, unlike Burt (1964) and Pindyck (1984), both of whom specify a stochastic soil evolution equation, non-stochastic production revenues and risk neutral farmer preferences, we have specified a non-stochastic soil evolution equation, a stochastic land price, stochastic revenues, and risk averse preferences; in this way, we are focusing our analysis on the effects of revenue risk and risk

---

[7]In short time intervals, production revenues may also be sensitive to an idiosyncratic random shock other than ε (e.g., transitory weather). However, for the medium to long run modelled here, these shocks will add up to a total shock which is likely to be small, to have little variability, and thus, to be unimportant. For this reason, as well as for the sake of analytical simplicity, we do not incorporate a third production-revenue-specific random shock in our analysis.

aversion for optimal farmer choices. Despite our focus on production and land risk, we should point out that our land price function in (5) does actually allow for some risk in soil evolution; in particular, the random shocks θ and ε can be interpreted as capturing random changes in both the soil base and the valuation of that base.

Fourth, in practice, agricultural production choices involve multiple inputs and multiple outputs, choices which may have complicated interactions in the determination of both soil evolution and the environmental impact of production. For simplicity here, we follow the soil conservation literature by abstracting from these interactions and focusing on a single output/ cultivation choice (e.g., see Burt (1981), Colby (1985), LaFrance (1992), McConnell (1983)). Although crop portfolio and input mix choices will affect soil evolution, it is also clear that cultivation intensities and fallowing practices are monotonically related to short-run production revenues and have the soil depletion effects modelled here; see the survey of evidence to this effect in LaFrance (1992). Here, we wish to learn about the effects of risk, risk aversion and insurance for general production activity level choices—and their attendant impact on soil depletion and environmental degradation—in a parsimonious model which abstracts from the multiple output/multiple input choices that merit study in their own right.

Fifth, in this paper, we do not explicitly model environmental externalities from agricultural production, such as adverse water quality impacts from irrigation run-off and percolation. Rather, we assume that such externalities are positively related to the farmer's output level and consider whether and when insurance programs may mitigate or exacerbate such environmental externalities by eliciting lower or higher output levels.

Finally, for simplicity, we model risk in current production revenues without distinguishing between risk in prices and outputs. This specification is valid, for example, if the current period output price is nonstochastic to the farmer, either generically or due to a government price stabilization program or due to variable-output forward contracting. Alternately, this specification is valid if production takes a pure multiplicative form (with $\alpha(Y) = 0$), in which case ε can be interpreted as the product of risky price and output.

Returning to our formal model, we can substitute for W in the farmer's utility function, V(W), using equations (3), (4) and (5). Taking expectations then yields the farmer's objective function and, in turn, the following farmer choice problem:

$$\max_{Y} J = EV[W_o + F(Y,\epsilon) - e(Y;X_1) + L(X_1 + G - g(Y), \varepsilon_L, \theta)] \quad (6)$$

where W_o is exogenous other income.[8] We assume that there is a finite solution to problem (6) in R^+. The first-order condition for this solution is:

$$J_Y = E\{V'(W)\, a\} = 0, \qquad a = [F_Y(\ ) - e_Y(\ ) - L_X\, g_Y] \quad (7)$$

Equation (7) equates (i) the utility gain from the marginal agricultural profits that are produced by higher output and (ii) the utility cost of the marginal land value loss that is produced by higher-output-induced soil depletion.

6.3 Risk and Insurance

The objective of this paper is to assess the effects of farmer insurance on farm output and social welfare. To make this assessment in the context of our simple model, we will consider two types of insurance; the first corresponds to the economist's

[8]In Ardila and Innes (1991), we consider a generalized version of this model which incorporates a choice of current consumption, and a corresponding choice of bond investment. Here, for simplicity, we abstract from this additional choice variable by fixing current consumption in the background. At the cost of complexity, the results presented here can be generalized to allow for a current consumption choice.

concept of pure insurance and will be studied in this section; the second corresponds to the "truncating" insurance form observed in practice and will be studied in the next section.

"Pure" insurance reduces the riskiness of the ε distribution in the classic sense of Rothschild and Stiglitz (1970); that is, we decompose e as follows (without loss of generality):

$$\varepsilon = \phi + \gamma h \tag{8}$$

where h is a mean-zero random variable, ϕ is the mean of ε, and γ is a mean-preserving spread parameter. An increase in γ represents an increase in risk (Rothschild and Stiglitz (1970)); equivalently, a reduction in γ represents an increase in "pure" insurance. In order to capture differential effects of an insurance program on the riskiness of current production revenues and the riskiness of land value, we will now make use of our ε_L distinction in the following way. At a given point (i.e., for a given value of γ), we define a fixed γ_o such that

$$\gamma_o + b\gamma = \gamma \Leftrightarrow \gamma_o = (1 - b)\gamma, \tag{9}$$

where b is a non-negative parameter. We further define ε_L as

$$\varepsilon_L = \phi + (\gamma_o + b\gamma)h, \tag{10}$$

so that $\varepsilon_L = \varepsilon$ at the initial γ. The parameter b thus captures the differential effect of insurance on production and land value risk. If b is small, then insurance has a small impact on land value risk relative to its impact on production revenue risk. Similarly, if b is large, insurance stabilizes the land value risk variable, ε_L, relatively more than production revenue risk. In what follows, we will be interested in the effects of marginal insurance, i.e., marginal decreases in γ, on output and welfare. From these marginal effects, we will be able to infer effects of discrete programs of pure insurance and the effects of corresponding discrete reductions in γ.[9]

Before proceeding with the analysis of "pure" insurance, a few comments are in order.

(1) Multiple-peril crop insurance is targeted at the stabilization of production revenues; together with forward/futures contracting, "pure" crop insurance will perfectly stabilize these revenues. However, even when a government crop insurance program is expected to persist in the indefinite future, it will not be capable of perfectly stabilizing land values. Random changes in output price distributions and production technologies insure that land price risk will persist, despite crop insurance. Therefore, in the context of our analysis, a pure crop insurance program is represented by a small b (b < 1).

---

[9]The careful reader will notice that, once γ_o has been defined in (9) (for a given initial $\gamma = \gamma_I$), a change in γ from its initial level will imply that ε_L (as defined in (10)) will no longer equal ε. However, once γ has changed from γ_I to its new level γ_N, the land-price relation can be redefined as a linear function of ε (without loss of generality). γ_o can then be redefined according to (9) and ε_L according to (10), with ε_L now equal to ε as before. The process of differentiating optimal outcomes with respect to γ (given b) can then proceed as before. In other words, for each γ, there is a γ_o and a linear land-price relation (as in (5)) such that $\varepsilon_L = \varepsilon$ and (10) is satisfied. Effects of discrete changes in γ can thus be inferred from effects of marginal changes in γ, given $\varepsilon_L = \varepsilon$ and ε_L as defined in (10). The specific transformation of the land-price relation that achieves this objective, given a change in γ from its initial level γ_I to its new level γ_N, is as follows:

$$\beta_o^N(X_2) = \beta_o^I + \phi \ell_I(X_2)(1 - \{[(1 - b)\gamma_I + b\gamma_N]/\gamma_N\})$$

$$\ell_N(X_2) = (b + \{(1 - b)\gamma_I/\gamma_N\})\ell_I(X_2)$$

where I and N index initial and new function values.

(2) In this analysis, we assume that the government is risk neutral and that some farmers are risk averse.[10] In view of this assumption, we will be able to infer relationships between output choices that are actually made by farmers and those that would be made by farmers that obtain complete—and hence, "first best"—insurance; these relationships can be inferred from the comparison of the output choices that are made under risk aversion and risk neutrality, respectively.

(3) While we do not present a complete welfare-theoretic analysis of optimal policy in this paper, we are interested in studying the effects of risk aversion and crop insurance on the welfare costs of two potential market failures. The first market failure in our model arises from the absence of complete contingent claim (insurance) markets. Due to incomplete risk markets, government insurance can potentially elicit trades of state-contingent incomes (between farmers and the government) that is welfare-increasing. However, inasmuch as the government insurance programs considered here need not replicate a complete-market equilibrium (i.e., they need not provide complete insurance), they may or may not elicit farmer output choices that are closer to the choices that would be made under complete insurance. It is arguably of some interest to know not only whether some risk-sharing gains are achieved by insurance programs but also whether or not the programs push farmer behavior toward outcomes that would be achieved with optimal insurance. Both of these issues are considered here. The latter question will be addressed by determining whether the insurance programs under consideration lead to output choices that are closer—or less close—to the choices that would be made under

[10]In the U.S., the assumption of a risk neutral government can be motivated by the extent of diversification in the government's "portfolio" of income bearing assets and the low level of systematic risk in farm-level returns that has been estimated in a number of studies (e.g., see Bjornson and Innes (1992) and the references therein). For developing countries, this assumption may be viewed as a modeling abstraction that captures the greater extent of risk aversion that prevails among farmers relative to the more-diversified government.

farmer risk neutrality. To address the former question, we structure both types of insurance so as to yield the government, which is the presumed insurance carrier, an expected payoff of zero. Therefore, given government risk neutrality, an insurance program will achieve risk-sharing gains if and only if farmer utility rises with the inception of the program.

The second potential market failure in our model concerns adverse environmental impacts/externalities associated with farmer output and, in addition, social value that might be attached to the soil stock (due to long-term government food security objectives, for example), but is not internalized by farmers. Both of these phenomena imply potential social costs of farmer output that are external to the farmers' optimization calculus. In what follows, we will be interested in determining whether or not the insurance programs considered here will serve to mitigate—or exacerbate—the divergence between private and social objectives implied by these externalities.

We should note at the outset that our examination of these welfare issues is a preliminary and incomplete one. In order to keep the analysis tractable and focused, we make a number of simplifying assumptions which abstract from potentially important analytical and policy issues. In addition to those mentioned earlier, these assumptions/abstractions include the following:

(i) Potential effects of insurance programs on consumers are not studied.[11]

(ii) We proceed in most of the analysis under the assumption that the insurance programs under consideration do not affect the land value relation directly (e.g., γ and $\overline{\varepsilon}$ are not explicit arguments of the land value function L( )). In interpreting many of our results, we also implicitly assume that the ex-post land value reflects the value that would prevail if all farmers were risk neutral. This second assumption implies that, in characterizing differences

[11]When output price is a function of production by potentially insured farmers (as in a closed agricultural market), then the effects of insurance programs on consumers may be important and rather complex (e.g., see Innes (1990a)).

between farmer choices and those that a "social planner" would make, we need not be concerned with differences between the land value relation that confronts farmers and that which the social planner would employ in his optimization problem. Both of these assumptions are valid if (i) there are a few risk neutral farmers in the population of land demanders, (ii) these risk neutral farmers have per-unit-land production functions that are invariant to the total land that they cultivate (i.e., they have constant returns to scale production technologies), and (iii) future production revenue expectations are unaffected by current insurance programs. Later in the paper (Section V), we consider prospective implications of an explicit land value dependence on government insurance.

Turning now to the formal analysis, the following preliminary result is useful:

Definition: Let $z = (f'(\ ) - \ell'(\ )\ g_Y(\ ))$. If z is positive at the optimum, then we will say that there is production risk dominance (PRD). If z is negative at the optimum, we will say that there is land risk dominance (LRD).

Proposition 1: When there is production (land) risk dominance, a risk averse farmer will produce less (more) than he would if he were risk neutral.

Proposition 1 indicates that, under land risk dominance (LRD), a standard result from the literature on production under uncertainty—that risk aversion leads to lower output—can be reversed. The intuition for this reversal is straightforward. Under LRD, investment in the soil stock "asset" is more risky than investment in the output "asset." Moreover, a risk averse farmer will invest less in the relatively more risky asset than will a risk neutral farmer (ceteris paribus). Thus, under LRD, risk aversion will lead to relatively less investment in soil, disinvestment which is achieved with a higher output level.

Since soil is the more risky asset under LRD, it is not surprising that an increase in risk can lead to an increase in output in our model. Under LRD, a pure increase in risk—i.e., an increase in γ with b set equal to one—will lead the risk averse farmer to shift further away from the relatively more risky asset, soil, toward the relatively less risky "asset," output.

Proposition 2: (A) Suppose there is PRD ($z > 0$) and $b \leq 1$. Then $dY^*/d\gamma < 0$. (B) Suppose there is LRD ($z < 0$) and $b \geq 1$. Then $dY^*/d\gamma > 0$. (C) For any $b < 1$, there exist cases of LRD ($z < 0$) such that $dY^*/d\gamma < 0$. (D) For any $b > 1$, there exist cases of PRD ($z > 0$) such that $dY^*/d\gamma > 0$.

Proposition 3: With any $b > 0$, an increase (decrease) in γ reduces (increases) farmer utility, E(V). Thus, regardless of b, a "pure" insurance program yields risk-sharing benefits.

Propositions 2 and 3 offer some central results of this paper. To interpret these results, first recall that, for a given reduction in γ, the parameter b indicates the extent to which production revenues are stabilized relative to land value. For expositional purposes, we will therefore refer to an insurance program with $b \leq 1$ as production-stabilizing and one with $b \geq 1$ as land-value-stabilizing. Using this terminology, the four cases in Proposition 2 have the following interpretation:

(A) When there is PRD, a production-stabilizing insurance program will lead to an increase in output.

(B) When there is LRD, a land-value-stabilizing insurance program will lead to a decrease in output.

(C) There are cases such that LRD holds and a production-stabilizing insurance program will lead to an increase in output.

(D) There are cases such that PRD holds and a land-value-stabilizing insurance program will lead to a decrease in output.

Now consider Proposition 2 (C). As noted earlier, pure crop insurance is a production- stabilizing type of program. Thus, for the cases described in part (C), crop insurance will lead to an increase in output. Moreover, since LRD prevails in these cases, the insurance program pushes farmers further away from their complete-insurance output level (Proposition 1), as well as exacerbating any environmental externalities created by agricultural production.

In contrast, a land-value-stabilizing insurance program will elicit an output decrease in these LRD cases (Proposition 2 (B)), thus pushing farmers closer to their complete-insurance output level and also mitigating environmental externalities.

Overall, Proposition 2 indicates that land-value-stabilizing insurance programs are likely to yield environmental benefits, while production-stabilizing insurance programs are likely to have environmental costs. When there is LRD, Proposition 2 also indicates that land-value- stabilizing insurance programs will tend to mitigate risk-induced over-production in agriculture. However, both types of insurance programs yield risk-sharing gains (Proposition 3).

Given the importance of the distinction between LRD and PRD cases, it is of some interest to determine which of these cases is relevant empirically. Although a complete treatment of this issue is beyond the scope of this paper, a cursory examination of returns received by U.S. farm operators suggests that the LRD case may be the more plausible one in U.S. agriculture. For the 1950-1986 period, the standard deviation of annual real capital gains on U.S. farm assets, as a proportion of asset values, was five times that of farm income, as a proportion of asset values.[12] Thus, the results of Proposition 2 suggest that governmental stabilization of land-value risk, rather than production risk, may well be the more appropriate object of policy.

6.4 "Truncating" Insurance

Traditionally, insurance does not take the "pure" form analyzed above, but instead provides some guarantee that losses will not be sustained beyond some given threshold. We now turn our attention to this traditional "truncating" type of insurance, which we model as follows:

First, we will distinguish between truncating production (or crop) insurance and truncating land insurance. With complete

[12]The data for this calculation are contained on page 5 of the Agricultural Finance Databook, June 1987, Board of Governors of the Federal Reserve System.

truncating production/crop insurance, a threshold level of ε, $\bar{\varepsilon}$, is defined and the farmer is insured production revenues of $F(Y, \bar{\varepsilon})$ when ε falls below $\bar{\varepsilon}$; when ε is above $\bar{\varepsilon}$, the farmer simply obtains his realized production revenues, $F(Y, \varepsilon)$. Similarly, with complete land insurance, a threshold $\bar{\varepsilon}$ is defined and the farmer is insured a land value of $L(X_2, \bar{\varepsilon}, \theta)$ when ε falls below $\bar{\varepsilon}$.[13] In each case, a nonstochastic insurance premium is levied on the farmer so that the expected cost of the insurance to the government is zero; in the case of production insurance, the premium is thus

$$\pi_F = f(Y)\, Q(\bar{\varepsilon}), \qquad Q(\bar{\varepsilon}) = \int_0^{\bar{\varepsilon}} (\bar{\varepsilon} - \varepsilon) p(\varepsilon)\, d\varepsilon$$

where $p(\varepsilon)$ is the marginal probability of ε; similarly, in the case of complete land insurance, the premium is

$$\pi_L = \ell(X_2)\, Q(\bar{\varepsilon})$$

Second, it is useful to consider combinations of partial truncating production and land insurance by defining two parameters, $b_i \in [0, 1]$, $i = F, L$, where b_F (b_L) represents the proportion of production risk (land risk) that is insured. For a given $\bar{\varepsilon}$ and (b_F, b_L) pair, the farmer obtains the following net ex-post government payment under combined truncating insurance (CTI):

$$[b_F\, f(Y) + b_L\, \ell(X_2)]\, q(\varepsilon, \bar{\varepsilon}) \tag{11}$$

where $q(\varepsilon, \bar{\varepsilon}) = \max(\bar{\varepsilon} - \varepsilon, 0) - Q(\bar{\varepsilon})$. The farmer's ex-post wealth level under this insurance program is then:

$$W = (W_o + \alpha(Y) + \beta_o(X_2) + \beta_1\theta - e(Y;X_1)) + (f(Y)$$

---

[13]We assume for simplicity that truncating land insurance applies only to ε-risk and does not incorporate effects of the variable θ. This assumption is valid when either (i) θ is degenerate (non-random) or (ii) θ and its effects on land value are observable and not the intended objects of agricultural insurance.

$$+ \ell(X_2))\varepsilon + q(\varepsilon,\bar{\varepsilon})[b_F f(Y) + b_L \ell(X_2)]$$

Notice that CTI encompasses a wide spectrum of special cases, including complete production insurance ($b_F = 1$, $b_L = 0$), complete land insurance ($b_F = 0$, $b_L = 1$), partial production insurance ($b_F < 1$, $b_L = 0$), partial land insurance ($b_F = 0$, $b_L < 1$), and complete joint insurance ($b_F = 1$, $b_L = 1$).

To determine the effects of combined truncating insurance (CTI) on output, we can totally differentiate the first order condition analog to (7) with respect to Y and the insurance threshold parameter $\bar{\varepsilon}$ to obtain:

$$dY^*/d\bar{\varepsilon} \overset{S}{=} E\{V''a^*(\partial q(\varepsilon, \bar{\varepsilon})/\partial\bar{\varepsilon})\}\ (f(Y)b_F + \ell(X_2)b_L) + E\{V'(\partial q(\varepsilon, \bar{\varepsilon})/\partial\bar{\varepsilon})\}\ (b_F f'(Y) - b_L \ell'(X_2)g_Y(\ )) \quad (12)$$

where "$\overset{S}{=}$" denotes "equals in sign,"

$$a^* = a + q(\varepsilon, \bar{\varepsilon})(b_F f'(Y) - b_L \ell'(X_2)g_Y(\ )) \quad (13)$$

$$a = \alpha'(Y) + f'(Y)\,\varepsilon - e_Y(\ ) - (\beta_o'(\ ) + \ell'(\ )\,\varepsilon)\, g_Y(\ ) \quad (14)$$

$$\partial q(\varepsilon,\bar{\varepsilon})/\partial\bar{\varepsilon}) = \begin{cases} 1 - Q'(\bar{\varepsilon}) = 1 - P(\bar{\varepsilon}) \text{ when } \varepsilon < \bar{\varepsilon} \\ -Q'(\bar{\varepsilon}) = -P(\bar{\varepsilon}) \text{ when } \varepsilon \geq \bar{\varepsilon} \end{cases} \quad (15)$$

and $P(\varepsilon)$ is the probability distribution function for ε.

Lemma 1: (i) $E\{V'\ (\partial q(\varepsilon, \bar{\varepsilon})/\partial\bar{\varepsilon})\} > 0$. (ii) Define ε^* such that $a(\varepsilon^*) = 0$ and $z' = ((1 - b_F)f'(Y) - (1 - b_L)\ell'(X_2)g_Y(\ ))$. Suppose $\bar{\varepsilon} \leq \varepsilon^*$. Then: (a) If $z' \geq 0$ and PRD holds, then $E\{V''\ a^*\ (\partial q(\varepsilon, \bar{\varepsilon})/\partial\bar{\varepsilon})\} > 0$; and (b) if $z' \leq 0$ and LRD holds, then $E\{V''\ a^*\ (\partial q(\varepsilon, \bar{\varepsilon})/\partial\bar{\varepsilon})\} < 0$.

Proposition 4: Suppose $\bar{\varepsilon} \leq \varepsilon^*$: $a(\varepsilon^*) = 0$. Then: (A) Joint truncating production and land insurance ($b_F = b_L \leq 1$) elicits an increase (decrease) in output under PRD (LRD). (B) Under LRD, there exists a positive $b_L^* \leq 1$ such that partial truncating land insurance ($b_L \leq b_L^*$, $b_F = 0$) elicits a decrease in output. (C) Under PRD, there exists a positive $b_F^* \leq 1$ such that partial truncating production insurance ($b_L = 0$, $b_F \leq b_F^*$) elicits an increase in output.

Proposition 5: Any truncating insurance program, whether for production revenue or land value or both, will increase farmer utility and thus yield risk-sharing benefits.

Propositions 4 and 5 extend some of our qualitative results on "pure" production- stabilizing and land-stabilizing insurance to "truncating" production and land insurance. Under LRD, for example, partial truncating land insurance leads to a decrease in output, thus mitigating environmental externalities and pushing farmers closer to their complete-insurance output level. Similarly, under PRD, partial truncating production insurance leads to an increase in output, again pushing farmers toward their complete-insurance output level but now at the price of exacerbating environmental externalities.

Despite prospective benefits of truncating land insurance, there are potentially non-trivial costs of implementing such an insurance program. For example, there are two reasons why the government would not want to compel a farmer to sell his property in order to obtain an insurance payment: (i) land sales are costly and (ii) it will often be optimal for the farmer to keep the land, which implies that either the land sale would lead to an inefficient allocation of resources or that the land sale may be made a sham. However, if the land need not be sold, then valuation of the property for purposes of determining the appropriate insurance payment becomes problematic and the potential object of dispute. Perhaps simple rules can be designed to assess the soil attributes of a particular piece of land and the corresponding designation of comparable land that has been sold in the recent past; in this case, a clear-cut assessment of land value could be easily made and land insurance would not be costly to implement. However, if land valuation is rather costly, then alternatives to specific-land insurance merit consideration.

One such alternative is as follows. Suppose there is a region in which agricultural land values are highly correlated and which is sufficiently large that, in any given time interval (say one year), the land that has been sold has average quality equal to the average quality of all land in the region. For simplicity here, we will assume that land values are perfectly correlated within this region. Now consider "truncating" insurance on the regional land value, rather than on the value of specific land. We will call such insurance <u>truncating regional land insurance</u> (TRLI).[14]

To assess the effects of TRLI, note that, unlike the truncating land insurance analyzed above, TRLI will not directly affect the farmer's marginal incentives to augment his own soil. However, TRLI insurance will still affect the farmer's risk and, hence, his risk preferences. Therefore, output choices will change with TRLI.

To be more specific, note that TRLI will add the following amount to farmer wealth:

$$\ell(\bar{X}_2)\, q(\varepsilon, \bar{\varepsilon}) \tag{16}$$

where $\bar{X}_2$ is average regional land quality at time 2 and (16) implicitly incorporates our assumption of perfectly correlated regional land prices.[15] Thus, we have

$$dY^*_{TRLI}/d\bar{\varepsilon} \overset{S}{=} E\{V''\, a(\partial q(\varepsilon,\bar{\varepsilon})/\partial\bar{\varepsilon})\}\,\ell(\bar{X}_2) \tag{17}$$

where a is as defined in equation (14).

Proposition 6: Suppose $\bar{\varepsilon} < \varepsilon^*$: $a(\varepsilon^*) = 0$. Then TRLI will lead to a decrease (increase) in output when there is LRD (PRD).

---

[14]TRLI is analogous to the "area yield" crop insurance discussed in Miranda (1991), except that it applies to land values rather than production revenues.

[15]As in our analysis of truncating land insurance, equation (16) reflects an assumption that TRLI does not incorporate effects of the variable θ (see footnote 10).

TRLI also increases farmer utility and thereby yields risk-sharing benefits.

Thus, TRLI pushes farmers closer to their complete-insurance output levels; in the case of LRD, TRLI also mitigates environmental externalities from agricultural production.

6.5 Land Value Effects of Insurance

So far in our analysis, we have assumed that government insurance programs do not affect the land value relation. This assumption is based on two premises: (1) that there are some risk neutral land demanders with constant-returns-to-scale production technologies, and (2) that future output and input price expectations are unaffected by the insurance programs under consideration. This second condition is implausible in markets where the potentially insured farmers provide a large share of total output supply. The first premise may also be violated when a managerial input in production leads to diminishing returns to scale.

In this section, we briefly consider implications of a land value dependence on a "pure" insurance program, allowing both of the above premises to be violated. To do this with a minimum of complexity, we consider a market with many identical risk averse farmers and write the land value function with two additional arguments, per-farm output Y and the insurance parameter γ, neither of which affects land risk,

$$L = L(X_2, Y, \gamma, \varepsilon, \theta) = \beta_0(X_2, Y, \gamma) + \ell\,(X_2)\varepsilon + \beta_1\theta \quad (18)$$

The per-farm output argument captures the following land price effects: First note that our assumption of identical farmers implies that per-farm output is proportional to total output by our population of potentially insured farmers. Thus, the argument Y in L( ) captures the effects of aggregate current output on future land value. Moreover, it is expected—and will be assumed here—that higher levels of aggregate output lead to higher land values, ceteris paribus; that is, $L_Y > 0$. The reason for this relationship is that higher aggregate outputs imply lower soil stocks, which in turn lead to higher future production costs and output prices; from the individual farmer's point of view, the higher future output prices (or, more precisely, superior future output price probability distributions) imply a higher value of land that has a given soil stock. For similar reasons, we can assume that marginal soil value is non-decreasing in the Y argument, $L_{XY} \geq 0$.

Note that, since the Y argument in L( ) captures effects of aggregate farm output that is beyond the control of the individual farmer, the farmer does not consider the direct effect of Y on L( ) in choosing his output. Thus, the condition that defines the farmer's output choice remains equation (7).

The γ argument in (18) captures different land price effects of insurance. In view of the farmer utility gains that an insurance program elicits (Proposition 3), it is to be expected that these gains, to the extent that they persist in the future with future insurance programs, will be capitalized in land prices. Thus, we will assume that $L_\gamma \leq 0$ and $L_{X\gamma} \leq 0$ (i.e., higher risk/less insurance does not increase the value of land or marginal soil).

Incorporating the land value relation in (18), we can differentiate the first order condition (7) to obtain:

$$dY^*/d\gamma = -S^{-1} \{ A + E\{V''a\} L_\gamma - E\{V'\} L_{X\gamma} g_Y( ) \} \quad (19)$$

where

$$S = E\{V''a^2\} + E\{V' [F_{YY} - e_{YY} + L_{XX} g_Y^2 - L_X g_{YY}]\} + E\{V''a\}L_Y - E\{V'\}L_{XY}g_Y \quad (20)$$

and A is the term that determined the sign of $dY^*/d\gamma$ in Proposition 2, i.e., A has the same sign as $dY^*/d\gamma$ when the land value does not depend explicitly on Y and γ.

Consider now the terms that define S in equation (20). The first, second and last terms are negative due to assumptions made at the outset. Under LRD, which implies that $E\{V''a\}$ is negative (Ardila and Innes (1991)), the third term is also negative. Under PRD ($E\{V''a\} > 0$), the third term is positive. However, despite this ambiguity in the case of PRD, a negative sign for S is plausible for the following reason: In order for the initial equilibrium to be stable (in the sense that a small perturbation of Y from Y^* would induce farmers to adjust their output choice toward Y^*, not away from it), S must be negative.

Turning now to equation (19), a careful inspection leads to the following conclusions:

Proposition 7: Suppose the land value function satisfies (18) with $L_Y \geq 0$, $L_{XY} \geq 0$, $L_\gamma \leq 0$, and $L_{X\gamma} \leq 0$. Then: (A) If there is PRD, equilibrium stability ($S < 0$), $b \leq 1$, and a sufficiently small $L_{X\gamma}$ at the optimum, then $dY^*/d\gamma < 0$. (B) If there is LRD and $b \geq 1$, then $dY^*/d\gamma > 0$. (C) For any $b < 1$, there exist cases of LRD and small $L_{X\gamma}$ such that $dY^*/d\gamma < 0$. (D) For any $b > 1$, there exist cases of PRD such that $dY^*/d\gamma > 0$.

Proposition 7 implies that the conclusions of Proposition 2 extends to an environment in which there are plausible land value effects of insurance.

6.6 Summary and Conclusion

This paper studies the effects of two types of insurance programs on a farmer's output choice when the farmer confronts risk in both production revenues and land value. Since both soil depletion and environmental externalities are positively related to output, environmental effects of insurance programs are also inferred from the analysis.

The two types of insurance programs considered are (i) "pure" insurance which stabilizes a random variable that underlies risk in both revenues and land price, and (ii) traditional "truncating" insurance which guarantees that farmer revenues and/or land price do not fall below some specified threshold. For each type of insurance, the analysis considers programs which target risk-reduction in either production revenues or land price relatively more. In addition, program effects are distinguished for two cases, one in which production risk dominates land risk (PRD) and the other in which land risk dominates production risk (LRD). It is argued that the latter LRD case may be of considerable importance in practice.

A key insight of this analysis is that higher levels of risk will, under LRD in our model, lead to higher levels of output and, hence, greater environmental degradation from agricultural production.[16] Therefore, the public has a potential stake in reducing agricultural risk beyond any interest that it may have in obtaining risk-sharing benefits from such risk reduction.

With respect to government insurance regimes targeted at agricultural risk reduction, the paper shows that production-revenue-stabilizing insurance (such as pure crop insurance) can elicit an increase in farmer output, thus exacerbating environmental externalities, leading to further soil depletion, and in the case of LRD, pushing farmers away from their complete-insurance (and "first-best") output levels. In contrast, under LRD, land-value-stabilizing insurance elicits lower output, thus mitigating environmental externalities and pushing farmers closer to their complete-insurance output levels. These benefits of land-stabilizing

---

[16]A positive relationship between risk and output can also be derived in the absence of risk aversion when debt-type financial contracts are present. Debt contracts lead to a risk of foreclosure (and loss of land), potentially truncating the farmer's payoff distribution and thereby eliciting behavior akin to that of a risk-lover, despite risk-neutral preferences. Although such arguments are interesting and potentially important, they need to be preceded by an explanation for the debt contract forms and the inability of contracting parties to alter contract terms that distort farmer behavior. In essence, an hypothesized presence of debt contracts must be attributed to underlying market imperfections that need to be modelled before the effects of these contracts can be assessed (see, for example, Innes (1990b, 1991)). In this paper, we deliberately chose not to take on these issues by instead focusing on the implications of risk aversion alone for production choices; this approach assumes only the absence of contingent claim markets, an absence which has abundant motivation in both theory (e.g., see Newbery and Stiglitz (1981)) and practice.

"pure" insurance are also found to derive from "truncating" land value insurance, whether this insurance is tied to specific farmland or to a regional land value index.

Whether one is primarily concerned with preserving the environment or with reducing deadweight costs of incomplete agricultural insurance markets, these results suggest that (i) government insurance of agricultural land values, rather than production revenue alone, may be an appropriate object of policy, and (ii) government insurance of production revenues alone may have some unfortunate consequences.

6.7 Appendix

Proof of Proposition 1: The F.O.C. (7) can be written,

$$E\{V'a\} = E\{V'\}E\{a\} + \mathrm{Cov}(V', a) = 0 \qquad (A1)$$

Since a risk neutral farmer sets E(a) equal to zero, the Proposition will follow if the covariance in (A1) is negative (positive) under PRD (LRD). By independence of q and e, this covariance is

$$\mathrm{Cov}(V', a) = \mathrm{Cov}_\varepsilon(E_\theta(V'), a)$$

Further, by risk aversion, $dE_\theta(V')/d\varepsilon < 0$, while $da/d\varepsilon = z > (<) 0$ under PRD (LRD). Thus, $\mathrm{Cov}(V', a) < (>) 0$ under PRD (LRD). QED.

Proof of Proposition 2: Totally differentiating condition (7) and using the second order condition for problem (6), we have:

$$\begin{aligned} dY^*/d\gamma &\overset{S}{=} E\{V''(dW/d\gamma)a\} + E\{V'(da/d\gamma)\} \\ &= E\{V''ah\}\, q + E\{V'h\}\, z' \end{aligned} \qquad (A2)$$

where "$\overset{S}{=}$" denotes "equals in sign,"

$$q = f(Y) + \ell(X_2)\, b > 0$$

$$z' = f'(Y) - \ell'(X_2)\, b\, g_Y(\ ) = z + (1 - b)\, \ell'(X_2)\, g_Y(\ )$$

Now note that $z' > 0$ in case (A) of the Proposition, $z' < 0$ in case (B), and $E\{V'h\} = Cov(V', h) < 0$. Thus, to prove parts (A) and (B), it suffices to show that

(I) $E\{V''ah\} > (<)\ 0$ under PRD (LRD).

Further, to prove parts (C) and (D), it suffices to show that

(II) $dY^*/d\gamma \overset{S}{\rightarrow} (b - 1)$ as $z \rightarrow 0$

where "$\overset{S}{\rightarrow}$" denotes "converges in sign."

To derive (I), rewrite $E\{V''ah\}$ by substituting for

$$h = (\varepsilon - \phi)/\gamma = \frac{1}{\gamma z}\{z\varepsilon - k + k - \phi z\},$$

and $a = z\varepsilon - k$, where $k \equiv e_Y + (\beta'_0 g_Y - \alpha')$. We then obtain:

$$E\{V''ah\} = [E\{V''(z\varepsilon - k)^2\} / \gamma z] + [E\{V''a\}(k - \phi z) / \gamma z] \quad \text{(A3)}$$

The first term in (A3) is equal in sign to $(-z)$, which is negative (positive) under production (land) risk dominance. To sign the second term, note that the first order condition (7) implies

$$(k - \phi z) = E\{V'h\} z\gamma / E\{V'\},$$

which is negative (positive) under production (land) risk dominance. Moreover, Ardila and Innes (1991) show that $E\{V''a\}$ and z are of the same sign. Thus, condition (I) follows from (A3).

To derive (II), note that as z goes to zero, $a = z\varepsilon - k$ goes to zero (from (7)), which implies that $E\{V''ah\}$ goes to zero. Thus,

$$dY^*/d\gamma \overset{S}{\rightarrow} E\{V'h\}\,(1 - b)\,\ell'(X_2)\, g_Y(\ ) \quad \text{as } z \rightarrow 0 \quad \text{(A4)}$$

Since the right-hand-side of (A4) has the same sign as (b - 1), condition (II) follows. QED.

Proof of Proposition 3: Differentiating E{V} yields:

$$dE\{V\}/d\gamma = E\{V'(\ )\ (dW/d\gamma)\} = E\{V'(\ )\}\ q > 0,$$

where q is as defined in (A2). QED.

Proof of Lemma 1: (i) Expanding the term of interest (using (15)),

$$E\{V'\ (\partial q(\varepsilon, \overline{\varepsilon})/\partial\overline{\varepsilon})\} = P(\overline{\varepsilon})\ \{E_{\varepsilon<\overline{\varepsilon}})\ \{E_\theta V'\} - E_\varepsilon\{E_\theta V'\}\} \quad (A5)$$

where $E_{\varepsilon<\overline{\varepsilon}}$ is the expectation over ε, conditional on $\varepsilon<\overline{\varepsilon}$. By risk aversion, $E_\theta(V')$ declines with ε, which implies that the right-hand-side of (A5) is positive.

(ii) (a) Expanding the term of interest as in (A5),

$$E\{V''\ a^*\ (\partial q(\varepsilon, \overline{\varepsilon})/\partial\overline{\varepsilon})\} = P(\overline{\varepsilon})\ \{E_{\varepsilon<\overline{\varepsilon}})\ \{E_\theta(V'')a^*\} - E_\varepsilon\{E_\theta(V'')a^*\}\} \quad (A6)$$

For the case of PRD, $z' \geq 0$, and $\overline{\varepsilon} < \varepsilon^*$, the graph of $E_\theta(V'')a^*$ is as depicted in Figure A1. In particular, this graph has the following properties:

(1) At $\varepsilon = \varepsilon^*$, $E_\theta(V'')a^* = 0$ (by the definition of ε^*, since $a = a^*$ at ε^*).

(2) For all $\varepsilon > \varepsilon^*$, $a^*(\varepsilon) = a(\varepsilon) > 0$ since $a(\varepsilon^*) = 0$ and $da/d\varepsilon > 0$ by PRD; therefore, since $E_\theta(V'') < 0$, $E_\theta(V'')a^* < 0$ for all $\varepsilon > \varepsilon^*$.

(3) For all $\varepsilon < \varepsilon^*$, $a^*(\varepsilon) < 0$ since $a^*(\varepsilon^*) = a(\varepsilon^*) = 0$ and $da^*/d\varepsilon = da/d\varepsilon > 0$ for all $\varepsilon \in [\overline{\varepsilon}, \varepsilon^*)$ and $da^*/d\varepsilon = z' \geq 0$ for all $\varepsilon < \overline{\varepsilon}$. Therefore, for all $\varepsilon < \varepsilon^*$, $E_\theta(V'')a^*$ is decreasing in ε; formally,

$$d(E_\theta(V'')a^*)/d\varepsilon = [E_\theta(V''')(dW/d\varepsilon)a^*] + [E_\theta(V'')(da^*/d\varepsilon)] < 0$$

where the inequality follows from (I) $E_\theta(V''') > 0$ (due to DARA preferences), (II) $dW/d\varepsilon = f + \ell > 0$ for $\varepsilon \in (\overline{\varepsilon}, \varepsilon^*)$ and $dW/d\varepsilon =$

$f(1 - b_F) + \ell(1 - b_L) > 0$ for $\varepsilon < \bar{\varepsilon}$, (III) $E_\theta(V'') < 0$ (risk aversion), and (IV) $da^*/d\varepsilon \geq 0$ for all $\varepsilon < \varepsilon^*$.

Properties (1) - (3) of the graph of $E_\theta(V'')a^*$ imply that, for any $\varepsilon_0 < \bar{\varepsilon}$ and any $\varepsilon_1 > \bar{\varepsilon}$, $E_\theta(V'')a^* |_{\varepsilon_0} > E_\theta(V'')a^* |_{\varepsilon_1}$, which implies in turn that the right-hand expression in (A6) is positive.

(b) For the case of LRD, $z' \leq 0$, and $\bar{\varepsilon} < \varepsilon^*$, the graph of $E_\theta(V'')a^*$ is as depicted in Figure A2. Therefore, following arguments analogous to those given above, the right-hand expression in (A6) is negative for this case. QED.

Proof of Proposition 4: (A) With $b_L = b_F = b$, wither $z' = 0$ (when $b = 1$) or $z' = z$. Thus, from Lemma 1(ii), PRD (LRD) implies that the first term in (11) is positive (negative). Further, from Lemma 1(i), the second term in (11) has the same sign as $z = f' - \ell' g_Y$ when $b_L = b_F$. Hence, since z is positive (negative) under PRD (LRD), part (A) of the Proposition follows.

(B) Under LRD ($z < 0$), there is a $\delta^* < 1$ such that $f'(Y) - \delta^* \ell'(X_2)\, g_Y(\,) = 0$. Now define $b_L^* = 1 - \delta^*$. Then, with $b_F = 0$ and $b_L \leq b_L^*$,

$$z' = f'(1 - b_F) - l' g_Y(1 - b_L) \leq f' - \ell' g_Y(1 - b_L^*) = 0$$

Therefore, from Lemma 1(ii), the first term in (11) is negative and, since $b_F = 0$, the second term is also negative from Lemma 1(i). Analogous arguments apply to part (C). QED.

Proof of Proposition 5: Differentiating E(V) with respect to $\bar{\varepsilon}$, we obtain:

$$dE(V)/d\bar{\varepsilon} = E\{V'(\,)(\partial q(\varepsilon,\bar{\varepsilon})/\partial\bar{\varepsilon})\}\ (f(Y)b_F + \ell(X_2)\, b_L) > 0$$

where the inequality follows from Lemma 1(i). QED.

Proof of Proposition 6: The bracketed right-hand expression in (17) can be written as in (A6) with "a" replacing "a*". Further, following arguments analogous to those in the proof of Lemma 1(ii), the graph of $E_\theta(V'')a$ is as depicted in Figure A1 (A2) under PRD (LRD), which implies that the right-hand-side of (17) is positive (negative) under PRD (LRD).

The second part of the Proposition is derived by differentiating E{V} with respect to $\bar{\varepsilon}$, and noting that the derivative is positive due to Lemma 1(i). QED.

Proof of Proposition 7: (A) From Proposition 2(A), A in (19) is negative here. Further, given PRD, $E\{V''a\} > 0$ (Ardila and Innes (1991)), so that $E\{V''a\}L_\gamma < 0$. Thus, since $S < 0$, $dY^*/d\gamma$ in (19) is negative when $L_{X\gamma}$ is sufficiently small. (B) From Proposition 2(B), A in (19) is positive here. Further, given LRD, $E\{V''a\}L_\gamma > 0$ and $S < 0$. Thus, since $-E\{V'\}L_{X\gamma}g_Y > 0$, $dY^*/d\gamma$ in (19) is positive. (C) From Proposition 2(C), A is negative as z approaches zero here. a—and hence, $E\{V''a\}$—also goes to zero when z goes to zero. Thus, when $L_{X\gamma}$ is sufficiently small (e.g., $L_{X\gamma} = 0$) and $b < 1$, the derivative in (19) is negative as z goes to zero. (D) This part follows from the same logic as part (C). QED.

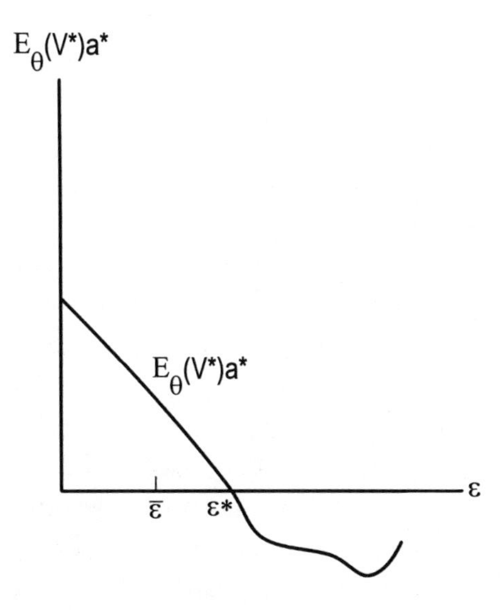

Figure A1

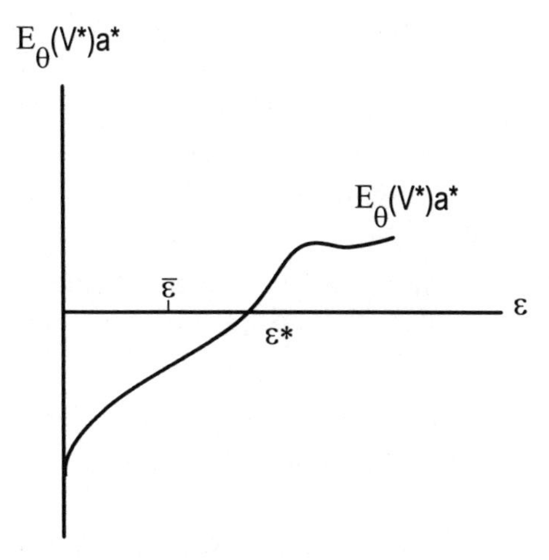

Figure A2

References

Ardila S, Innes R (1991) Risk, risk aversion and on-farm soil depletion. Univ of Arizona, working paper

Arzac E (1989) Income insurance with uncertain output. Int Econom Rev 30:561-570

Bjornson B, Innes R (1992) Another look at returns to agricultural and nonagricultural assets. Amer J Agr Econ 74:109-119

Burt O (1964) Optimal resource use over time with an application to ground water. Management Science 11:80-93

Burt O (1981) Farm level economics of soil conservation in the palouse area of the northwest. Amer J Agr Econ 63:83-92

Chambers R (1989) Insurability and moral hazard in agricultural insurance markets. Amer J Agr Econ 71:604-616

Colby B (1985) Soil productivity and farmers' erosion control incentives. Western J Agr Econ 10:354-364

Gardner K, Barrows R (1985) The impact of soil conservation investments on land prices. Amer J Agr Econ 67:943-947

Innes R (1990a) Government target price intervention in economies with incomplete markets. Quart J Econ 106:1035-1052

Innes R (1990b) Imperfect information and the theory of government intervention in farm credit markets. Amer J Agr Econ 72:761-768

Innes R (1991) Investment and government intervention in credit markets when there is asymmetric information. J Public Econ 46:347-381

Just R, Pope R (1987) Stochastic specification of production functions and economic implications. J Econometrics 7:67-86

King D, Sinden J (1988) Influence of soil conservation on farm land values. Land Econ 64:242-255

Kramer R, McSweeney W, Stavros R (1983) Soil conservation with uncertain revenues and input supplies. Amer J Agr Econ 65:694-702

LaFrance J (1992) A dynamic analysis of land degradation and prices. Aust J Agr Econ forthcoming

McConnell K (1983) An economic model of soil conservation. Amer J Agr Econ 64:83-89

McSweeney W, Kramer R (1986) Soil conservation with uncertain revenues and input supplies: reply. Amer J Agr Econ 68:361-363

Miranda M (1991) Area yield crop insurance reconsidered. Amer J Agr Econ 73:233-242

Miranowski J, Hammes B (1984) Implicit prices of soil characteristics for farmland in Iowa. Amer J Agr Econ 66:745-749

Newbery D, Stiglitz J (1981) The theory of commodity price stabilization. Oxford Univ Press

Pindyck R (1984) Uncertainty in the theory of renewable resource markets. Rev Econ Stud 51:289-303

Quiggin J (1992) The optimal design of crop insurance. This book, chapter 5.

Rothschild M, Stiglitz J (1970) Increasing risk I: a definition. J Econ Theory 2:225-243

Saha A, Innes R, Pope R (1991) Production and savings under uncertainty. Texas A&M Univ, working paper

Sandmo A (1971) On the theory of the competitive firm under price uncertainty. Amer Econ Rev 61:65-73

Shortle J, Stefanou S (1986) Soil conservation with uncertain revenues and input supplies: comment. Amer J Agri Econ 68:358-360

Chapter 7

Crop Insurance In The Context Of Canadian and U.S. Farm Programs

A. SCHMITZ[1], R.E. JUST[2], and H. FURTAN[3]

In the United States and Canada, as in most countries, there exist a number of farm programs. Some programs, such as crop insurance, are aimed at reducing the risk of crop failure while other programs are aimed at stabilizing commodity prices. In Canada, crop insurance is widely used by Prairie grain producers; in 1989, more than 70 percent of grain producers carried it. However, other programs are also commonplace on the Canadian Prairie. For example, in 1989, roughly 90 percent of the producers participated in the Western Grain Stabilization Act. Other programs benefiting all farmers included the Special Canadian Grains Program and the 1988 drought payment.

Use of crop insurance in the United States was highly varied with higher participation in the areas with greater relative yield variability. Participation rates in 1987, for example, were only 5.3 percent with the relative yield stability of Ohio soybean production but were as high as 70.6 percent in the highly volatile yield variability of Montana wheat production.[4] Other important farm programs which have, to varying extents, stabilizing and

---

[1]Department of Agricultural and Resource Economics, University of California, Berkeley, California, USA.

[2]Department of Agricultural and Resource Economics, University of Maryland, College Park, MD 20742, USA.

[3]Department of Agricultural and Resource Economics, University of Saskatchewan, Saskatoon, Saskatchewan, Canada.

[4]It should not be inferred from the low participation rates that crop insurance in the United States is a failure. In areas where yields are relatively stable, the objectives of crop insurance can be achieved with a low participation rate.

supporting influences also drew significant participation. These included the Conservation Reserve Program, which permitted farmers to draw a stable income payment on land taken out of production, as well as farm programs for specific commodities such as wheat, feed grains, rice, and cotton with their accompanying acreage reduction programs, deficiency payments, and farmer-owned reserves.

This paper examines crop insurance in both the United States and Canada and its interaction with other farm programs. A theoretical model of crop insurance is developed. Other farm income programs such as the Western Grain Stabilization Act in Canada and the target price-loan deficiency scheme in the United States are discussed. A theoretical model of how these programs interact with crop insurance is provided.

While this paper examines the interaction of crop insurance with the wide variety of farm programs that exist in Canada and the United States, the Canadian Western Grain Stabilization Act is more clearly a stabilization program than are many of the U.S. farm programs that are offered in conjunction with crop insurance. Accordingly, much of the analysis of program interaction is cast in the context of the Western Grain Stabilization Act. The first two sections of this paper describe the major government programs in Canada and the United States. Then several sections examine the implications of crop insurance and the interaction of crop insurance with other programs.

7.1 Canadian Grain Programs

A brief summary of the major Canadian programs follows. In the 1980s, there were large increases in transfers from governments to producers (Fig. 1). This was largely in response to droughts and low grain prices. Of major importance in the late 1980s were the Western Grain Transportation Act, crop insurance, the Special Canadian Grains Program, the Western Grain Stabilization Act, and drought assistance.

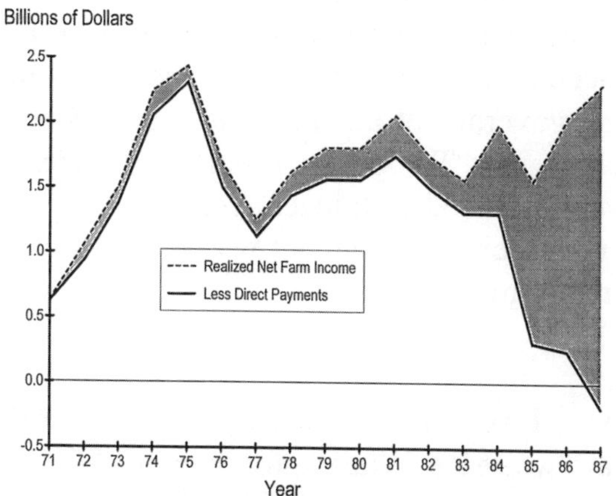

Fig. 1. Direct Government Payments and Realized Net Income, Prairie Region, 1971–1987

Source: Fulton, Rosaasen, and Schmitz

Over time, some programs have been eliminated and new ones have been introduced. The first section describes briefly those programs which were important historically but were no longer in place as of January 1, 1989. The second section deals with government programs currently in place.

7.2 Past Programs

7.2.1 Temporary Wheat Reserve Act. During the period 1955-1970, a major program was the Temporary Wheat Reserve Act. This program resulted in a major income transfer to Prairie farmers. During that period, Canada accumulated large grain stocks. Under the Temporary Wheat Reserve Act, farmers were partly compensated for on-farm storage.

7.2.2 Lower Inventory for Tomorrow. During the late 1960s, there was a glut in the world wheat market. Large wheat stocks had accumulated in both the United States and in Canada. The Canadian government attempted to reduce stocks through the Lower Inventory for Tomorrow program. Under this program, producers were provided with incentives to reduce wheat plantings. The cost to the federal government was $63 million. This amount was paid directly to farmers.

7.2.3 Two-Price Wheat. The federal government introduced a Two-Price Wheat program in 1967 to stabilize the price of wheat to Canadian producers and consumers. Under the program, if the Canadian wheat export price fell below a specified domestic floor price, the Canadian consumers paid the floor price—in effect, a subsidy to domestic producers. However, if the export price rose above a specified ceiling price, then consumers paid only the ceiling price. In this case, the subsidy flowed from producers to consumers. When the export price fell between the ceiling and floor price, then the domestic and export prices were equal; hence, no group would receive a subsidy. In the late 1960s and early 1970s, producers gained at the expense of consumers but in the late 1970s, consumers gained at the expense of producers.

7.2.4 Special Canadian Grains Program. In response to the drop in the U. S. loan rate, the Canadian government introduced the Special Canadian Grains Program in December 1986. A payout of $1 billion was made to Canadian grain and oilseed producers. The payments were made in 1987 in two installments which amounted to $300 million and $700 million, respectively. Payments were based on the acreage that farmers seeded to the designated crops in 1986, the regional crop insurance yield, and the relative price decline in each commodity due to the trade war between the United States and the European Economic Community. The crops covered under the program included wheat and durum, barley, oats, rye, mixed grains, corn, soybeans, canola, flax, and sunflower seeds. The maximum payment to any individual was $25,000.

Because of the continuing low prices, in December 1987, the government again announced the Special Canadian Grains Program. It totaled $1.1 billion, and the limit was set at $27,500 per producer. This amount was paid out in 1988. The basis of the payment was somewhat different than the previous program in that summer fallow acreage was considered along with additional crops.

7.2.5 Drought Payment. The 1988 Prairie grain crop was affected by one of the worst droughts in the history of Canada. In certain areas, yields were less than during the drought of the 1930s. In 1988, the government announced a drought-assistance program for Canadian farmers. The Prairie region received payments in excess of $500 million. Payments were made in April and May of 1989 and, again, in August and September of 1989. These payments were based on 1988 yields and were assessed on a township basis. In the payments, factors such as land quality were included. Everything else equal, the higher the land quality, the greater was the payout.

7.3 Existing Farm Programs

At least three major farm programs have been in place for several years and are still in place as of January 1, 1989. These are the crop insurance program, the Western Grain Stabilization Act, and the Western Grain Transportation Act.

7.3.1 Crop Insurance. Yields of crops grown on the Prairies have fluctuated dramatically as has grain quality. Farmers have always supported public crop insurance programs. The first crop insurance program was introduced in 1960. Crop insurance provides for losses caused by natural hazards such as frost, fire, floods, hail, insects, plant diseases, and drought. Crops covered include wheat and durum, oats, barley, flax, canola, rye, sunflowers, mustard, utility wheat, and canary seed. The program is funded by producer premiums which are matched by the federal government. However, each provincial government pays the administrative cost of the program.

Crop insurance programs differ among provinces. In Saskatchewan, the program offers farmers the choice of 60 percent or 70 percent coverage for their risk area and soil class. For example, in a given area, if the long term yield is 30 bushels per acre on summer fallow, then, if the farmer chooses 70 percent coverage, he or she is insured for 21 bushels per acre. Associated with yield levels are various prices which the producer can select. In 1989, for example, under crop insurance, the farmer could select one of three prices and the premiums varied accordingly. Many farmers chose the variable high-price option. Prior to 1989, there were essentially only two prices from which farmers could choose. These final prices were identical to the prices agreed to at the time the crop insurance was taken out. Thus, if in the spring of the year farmers took out the high price which, say, was $4.00 per bushel, they were automatically guaranteed $4.00 per bushel if crop insurance was collected in the fall of the year. The high price option was introduced in 1989. Which crop insurance premiums are used depend on many factors including soil type, type of crop grown, type of crop rotation, and the level of coverage selected. In addition, coverage levels increase and premiums decrease if the farmer has a high-performance record (that is, no claims or small claims).

The average annual federal government cost for crop insurance for the Prairie provinces over the five-year period 1981-82 to 1985-86 was $120 million. However, in both 1988 and 1989, the amounts far exceeded this level (Table 1). Because of the severe drought in 1988, the payout exceeded $500 million. Given the flooding conditions that existed in 1989, crop insurance payouts could easily approach $500 million.

Table 1. Cash receipts and program payouts

Year	Total cash receipts[a]	Total grain cash receipts[b]	Net farm income	Crop insurance payments	Western Grain Stabilization Act
			thousand dollars		
1978	2,504,017	1,872,521	881,568	28,000	67,900
1979	3,022,173	2,333,530	802,910	103,010	150,400
1980	3,302,927	2,545,816	648,391	135,334	0
1981	4,008,935	3,284,595	1,574,572	92,010	0
1982	4,027,555	3,248,666	1,108,735	82,043	0
1983	3,965,045	3,261,322	582,171	105,946	0
1984	4,362,779	3,537,927	343,582	171,752	125,940
1985	4,101,383	3,291,835	868,470	282,693	292,357
1986	4,138,312	3,222,890	1,390,485	289,849	477,144
1987	4,237,284	2,976,609	940,020	110,289	759,238
1988	4,438,463	2,944,182	207,712	263,825	378,496
1989	4,644,637	3,157,189	1,217,499	503,672	135,455

[a]For all commodities and all programs.

[b]For all grain and all grain programs.

Source: Statistics Canada 21-603.

Not all farmers use crop insurance. However, a record number of farmers participated in the crop insurance program in 1989. Those who do not are generally well financed and have landholdings scattered in different locations. Also, certain producers are in and out of the program depending on factors such as spring climatic conditions. Some producers only insure against possible hail losses through private companies. Over 80 percent of the producers in the Prairie region have some form of crop insurance. The percentage varies from year to year since, at low market prices, the coverage is less and, hence, a certain percentage of the farmers will not take out insurance.

The yield base for insurance is the 10-year moving average of the yield on each class of soil. The farmer can insure his crop at 60 or 70 percent of this yield base. What changes from year to year is the price at which these yields can be insured. For example, those farmers in 1989 who opted for the high price option would receive a price coverage of at least $1.00 per bushel higher than the maximum price coverage available during the 1988 season. The low price coverage in 1988 partly explains why the federal government responded to the severe drought by introducing a drought-assistance payment.

7.3.2 Western Grain Stabilization Act. The Western Grain Stabilization Act was introduced in 1976 to deal with large variations in the income of Prairie farmers. The program was put in place to avoid a repetition of the economic downturn of the late 1960s which was due to declining international grain prices and sales. The program is voluntary. Farmers who join the program contribute a percentage of their gross sales—up to a maximum of gross sales of $60,000—to the stabilization fund. The federal government also contributes to the fund (Table 2). Over time, the levy paid by farmers has increased from 1 percent of a maximum of $60,000 to 4 percent. The seven major grains grown on the prairies (wheat and durum, oats, barley, rye, flax, canola, and mustard) are eligible under the program. The one exception is grain fed to livestock on the farm where it is produced.

The average annual contribution by the federal government for the five-year period 1979-1983 was $107.6 million. Payouts from the program are triggered when the net cash flow from the seven grains grown in the Prairie region falls below 90 percent of the previous five-year average net cash flow. The payout to an individual producer is determined by the level of his/her levy for the current and previous two years. The above formula is designed to stabilize regional net cash flow. However, the income of the individual producer may not be stabilized. For instance, this would occur if the farmer had a poor crop, yet the Prairie region enjoyed a good crop and relatively high prices.

Since 1983-84, the Prairie grain economy has worsened significantly, largely because of sharply falling grain prices. This brought about a significant increase in payouts under the Western

Table 2. Cumulative stabilization account January 1, 1976 to July 31, 1988

Year	Levy		Interest	Payouts
	Producer	Government		
	million dollars			
1976	24.3	48.6	1.3	
1977	28.0	56.0	7.1	
1978	28.4	56.8	11.7	115
1979	43.4	86.8	9.2	253
1980	48.3	96.6	9.4	
1981	56.4	112.8	44.5	
1982	55.5	111.0	59.9	
1983	65.3	130.6	60.5	
1984[a]	26.8	62.5	52.5	223
1984-85	45.5	106.2	75.9	522
1985-86	29.9	89.7	35.4	859
1986-87	27.4	82.2	(18.5)	1,396
1987-88	118.4	177.6	(139.0)	693
TOTAL	597.6	1,217.5	209.9	4,061

[a]Covers January through July.

Source: Agriculture Canada, Western Grain Stabilization Annual Report, 1987-88.

Grain Stabilization Act. Farmers who contributed a maximum levy received roughly $25,000 under the Western Grain Stabilization Act for the calendar year 1987. In December 1987, the federal government put in an additional $750 million into the stabilization fund. This amount was still insufficient to cover the deficit created by the Western Grain Stabilization Act. As a result, the levies to producers were increased to 4 percent of a maximum of $60,000 of grain sales for a given year. Not all producers belong to the Western Grain Stabilization Act since it is a voluntary program.

However, for the four Prairie provinces, participation averaged 80 to 85 percent for the 1986-87 crop year and 88 to 91 percent for the 1987-89 crop years.

7.3.3 Western Grains Transportation Act. The Western Grains Transportation Act was passed in November 1983. Under the Act, the federal government provides railways with an annual payment of up to $658.6 million (plus an inflation index) to cover the transportation of eligible grain from Prairie shipping points to Thunder Bay, Churchill, Vancouver, and Prince Rupert's Land. The $658.6 million is referred to as the Crow Gap. It was an estimate of the shortfall in revenues experienced by the railways in moving grain at the statutory rates at the time the legislation was introduced. In years when exports are low, the payout under the Western Grains Transportation Act may be less than $658.6 million because of small volumes. The amount paid out each year is calculated on a dollar-per-tonne moving basis and varies with the distance to port.

The 1983 Act replaced what is known as the Crow Rate. The Crow's Nest Pass Agreement was introduced in 1897 as an instrument for economic development in Canada. The federal government provided a subsidy to the Canadian Pacific Railway Company for the construction of a railroad in southeastern British Columbia. The Canadian Pacific Railway Company agreed to reduce the freight rates on grain and flour moving eastward out of the Prairies and on certain settlers' effects moving into the west. The Crow Agreement was used as an instrument to integrate the Canadian economy, to encourage the development of the West, and to provide inexpensive food for the population of Central Canada. Under the Crow Agreement, the rate at which grains and oilseeds could be shipped out of Western Canada (known as the statutory rate) remains fixed. However, as time passed, the railways contended that those rates were too low to cover the cost of an adequate and efficient transportation system. In response, the federal government paid for branch-line maintenance while both the federal and provincial governments undertook the purchase of the railway cars. Because of rising costs for branch-line maintenance and the increased demands for policy change from the railways, livestock producers, processors, and selected grain groups the federal government enacted new legislation in 1983.

7.4 Net Transfers

The net transfer to Prairie producers from the selected government programs discussed above are given in Fig. 2. This covers the period 1953-1987. As the data indicate, for many years, the largest transfer from the government to the producers was through the Western Grains Transportation Act. However, in 1987, the Special Canadian Grains Program and the Western Grain Stabilization Act also became major farm programs. In addition, in the years following 1987, another special grains program and a drought program provided major transfers to producers.

It is interesting that during the development of Prairie agriculture, much of the debate centered on how commodities should be marketed rather than the extent to which the farmer's

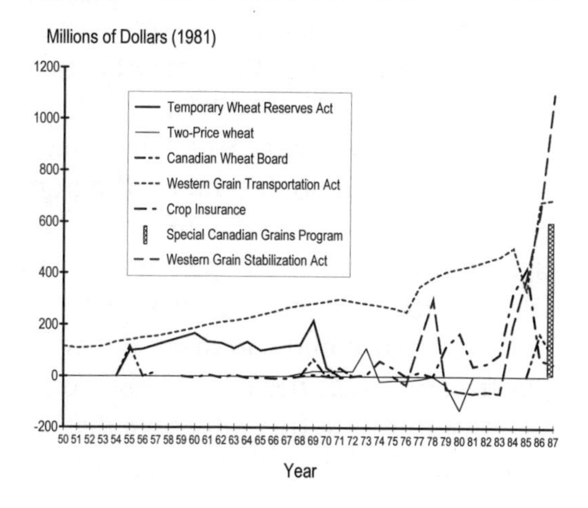

Fig. 2. Net Transfers to Prairie Producers under Selected Government Programs

Source: Fulton, Rosaasen, and Schmitz

income should be supported through farm programs. For most of the period prior to 1950, there were very few transfers to Prairie producers from either the provincial or federal governments. Prior to that time, the Crow Rate generally was not a subsidy to producers. No other program of any major significance was in place. Generally it is safe to assume that the cumulative total of government transfers to producers in the Prairie region from 1984 to 1989 will far exceed, in real terms, the cumulative total for all of the years prior to that period. During the latter part of the 1980s, droughts, coupled with the lowest real price of wheat in history, resulted in income transfers from the government to Prairie grain producers. Since crop insurance is the major focus of this paper, it is important to stress that crop insurance does not substitute for other programs in terms of income transfers to producers.[5]

The Western Grain Stabilization Act and crop insurance programs are complementary in stabilizing farm incomes. If a farmer has a crop failure and receives a payment, he can use this payment as part of the $60,000 criteria. Thus, crop insurance payments make up for the shortfall in production or revenue that form the basis for the net cash flow support program, the Western Grain Stabilization Act. Also, as Fig. 2 shows, you can have a large crop insurance payout and no Western Grain Stabilization Act payment (see, for example, 1979 to 1984).

In 1991 a major policy change went into effect for Canadian producers. The Gross Revenue Insurance Program was put in place, replacing crop insurance. Under this program, farmers are guaranteed a specific gross income per acre for a grain crop regardless of whether there is an income shortfall from the low market prices, yields, or a combination of the two. Under this program a producer can collect because of low commodity prices even if the farmer's yield was sufficiently high that a payout would not have been made under the previous crop insurance program.

---

[5]There may be an indirect link between crop insurance and Special Canadian Grains Program in terms of the amount of stubble crop.

The gross revenue insurance program, however, is administered through crop insurance and the yield data and the like used historically by crop insurance form the basis for the new program.

7.5 United States Grain Programs

We now consider farm programs in the United States. This section describes past and present U.S. agricultural policies, and the following section extends the analysis to the U.S. case.

7.6 Past Programs

Many U.S. farm programs are limited to specific crops: wheat, feed grains, cotton, rice, and sugar. The major policy instruments used with respect to these crops have consisted of price supports, target prices, deficiency payments, set-asides and acreage reduction programs, a soil bank, the Conservation Reserve Program, nonrecourse loans, farmer-owned reserves, and export enhancement programs.

7.6.1 The Agricultural Adjustment Acts of 1933 and 1938. The U.S. agricultural policy over the past five decades has its roots in the Agricultural Adjustment Acts of 1933 and 1938. The Agricultural Adjustment Acts were enacted originally to raise farm incomes and control production. Under the Agricultural Adjustment Acts, farmers were assigned allotments based on the average of past acreages. They could then reduce the area by a certain percentage of this allotment base and in return receive a cash payment. While the production control features of the 1933 Act were declared unconstitutional, the 1938 Act introduced nonrecourse loans, storage payments, direct payments, allotments, marketing quotas, and conservation incentives that remain in spirit in the current legislation.

7.6.2 Farm Policy in the Postwar Period. Various attempts to suspend marketing quotas or allotments over the years, and a reluctance to lower support prices from the 1910-1914 parity standards caused the government to acquire increasingly large stocks in the post-World War II period. This led to the Agricultural Act of 1956 and the establishment of the Soil Bank to withdraw farmland from production and help reduce the growing surpluses. In the 1960s, there was a switch toward voluntary government programs amidst chronic surpluses, renewed efforts to retire land from production, and massive food aid exports under Public Law 480.

7.7 Existing Farm Policies

The specific policy controls used in farm programs today were first introduced during the 1970s. These policy controls, for the most part, had a dual role of stabilizing farm income and commodity markets and supporting farm incomes.

7.7.1 The Agriculture and Consumer Protection Act of 1973. The Agriculture and Consumer Protection Act of 1973 introduced the target price concept whereby deficiency payments were made to farmers on the basis of the difference between the target price and the higher of the market and support prices. The goal of the target price system was to support income at a stable level without affecting the market price. Under this program, farmers could be relatively certain about the price they would receive for the bulk of their crop when target prices were above market prices (which occurred most of the time). The 1973 Act also introduced the disaster payments program. Under this program, participating farmers received payments as a percentage of target prices to cover losses due to natural causes which either prevented crop planting or resulted in low yields. Essentially this program offered participating farmers free crop insurance. It was the first farm bill to recognize and provide broad coverage to farmers against yield risk.

7.7.2 The Food and Agriculture Act of 1977. The Food and Agricultural Act of 1977 attempted to provide a further stabilizing market influence by introducing the farmer-owned reserves program. Under this program, farmers received loans and annual storage payments in return for agreements to store grain for three to five years unless the market price reached a specified release price. Unlike previous programs where the government acquires ownership of commodities at low prices, this program allows farmers to gain the benefits of future price rises by retaining ownership. The program attempted to provide farmers with the benefits of a price-band type of price stabilization policy while financing the associated storage cost incurred by the farmer. In practice, however, the program failed to provide this type of price stabilization because the price-support/release-price band was set so high that market prices virtually never reached the release levels while often reaching the support level. Optimistic price control levels made the program operate more as an income-support program.

7.7.3 The Agriculture and Food Act of 1981. The acreage reduction program was introduced under the Agriculture and Food Act of 1981. Under this program, land was diverted from the production of specified crops and the diverted land was put into an approved conservation use. The Secretary could adjust the parameters of the acreage reduction program to reduce acreage when prices were low and stocks were high, thus providing an endogenous market-stabilizing influence. Weak market conditions and overly optimistic target prices in the early 1980s, however, prevented the program from having a strong stabilizing influence. Rather, the acreage reduction program was operated at high levels simply to counteract the supply-inducing target prices.

7.7.4 The Food Security Act of 1985. With the Food Security Act of 1985, the target and supply prices were lowered to more realistic levels, and more flexible use of the acreage reduction program was possible. The 1985 Act also introduced a new Conservation Reserve Program whereby farmers could be paid a fixed annual rental payment for retiring part of their base acreages. As with most other programs, this program has both an income-supporting element in that it attempts to reduce the oversupply of certain crops and a stabilizing element in that it provides participating farmers with a fixed income source.

7.7.5 Crop Insurance. Federal crop insurance was introduced in the United States much earlier than in Canada. The Federal Crop Insurance Corporation (FCIC) was created in 1938 and has been in virtually continuous operation since. However, the level of operation over much of the period since has been modest. Both the set of covered crops and the number of counties in which insurance is offered has varied greatly. Initially, only insurance for wheat was offered. Participation levels have also been low throughout most of this period. Much of the low participation is explained by the need to set insurance parameters at levels that control losses when only the most risky farmers participate (adverse selection) and the competition of other farm programs such as the disaster payments program which offered free crop insurance to farmers under the Agriculture and Consumer Protection Act of 1973 and the Food and Agriculture Act of 1977. The relative importance of the disaster payments program is reflected in the fact that in the period 1974-1980 disaster payments were about three-and-a-half times as much as total FCIC indemnities.

Federal all-risk crop insurance became a more important instrument of agricultural policy with the Federal Crop Insurance Act of 1980. This program attempted to replace disaster payments by establishing a 30 percent subsidy for FCIC premiums, by expanding coverage to new counties and commodities, and by using the private insurance industry to promote sales. With this program, FCIC premiums which had not exceeded $50 million per year until 1974 and had barely exceeded $100 million by 1979 jumped to almost $600 million in 1982. Participation jumped from 9.6 percent in 1980 to 24.5 percent in 1988. Still, this level of participation is significantly lower than in Canada and lower than the program goal of 50 percent.

Again, an apparent explanation lies in the interaction of crop insurance with other agricultural legislation. On several occasions in the 1980s, Congress enacted ad hoc disaster bills in response to crop failures. As with the 1973 and 1977 farm bills, these programs offered farmers free crop insurance against major disasters. Although the availability of these programs could not be anticipated with certainty because they were enacted on an ad hoc basis as disasters arose, farmers were provided with a clear signal that the federal government would be likely to make major disaster payments. Disaster payments under these bills averaged $800 million annually in 1981-1988, and exceeded FCIC indemnities which averaged only about $600 million annually over the same period.

In hindsight, crop insurance has not replaced disaster payments and vice versa. A farmer can receive disaster payments, deficiency payments associated with target prices, and crop insurance payouts. Thus, the programs are not perfect substitutes as far as farmers are concerned. In a disaster year he can collect payments from all three sources.

7.7.6 An Overview of Existing Programs. This summary of U.S. agricultural policy is brief and omits a myriad of other more minor variations in the set of policy instruments and instrument levels. However, it shows that a number of the major policy controls—particularly those introduced in recent years—have a potential stabilizing as well as an income-supporting effect. The stabilizing effects are of particular interest because they can potentially interact with federal crop insurance, possibly reducing its benefits and the demand for it.

As in Canada, farmers have received sizable transfers from the government through these various farm programs. The growth in government transfers is illustrated in Fig. 3 which depicts U.S. net farm income and the portion of that income due to direct government payments. This figure parallels Fig. 1 which gives the Canadian experience. In the United States, as in Canada, the share of net farm income provided by direct government payments has been increasing over time. However, the relative extent of the increase has not been nearly as dramatic.

In examining Fig. 3, it is interesting to note that direct government payments have done little to stabilize farm income. When farm income declined sharply in 1974-1976 and again in 1980, direct payments declined rather than increased. When farm income had a major decline in 1983, direct payments increased sharply. However, they did not decline substantially in the following years when farm income rebounded. In fact, when farm income reached its all-time high in 1986, direct payments were also at their all-time high. Fig. 3 illustrates that, although U.S. farm programs have been increasingly designed to exert a stabilizing influence on agricultural commodity markets and farm incomes, they have tended to be used in practice to support incomes. Also, once the level of payment reaches a high level in a bad year, it is difficult to reduce government payments in good years.

7.7.7 Comparison of U.S. and Canadian Programs. A number of comparisons can be made between the U.S. and Canadian farm programs. First, in both countries the other major programs are not perfect substitutes for crop insurance; in fact, they appear to be complementary. Second, in both countries the purpose of most other programs is largely to support farm prices and incomes while the purpose of crop insurance is stabilization or risk reduction. In the United States, prices are supported through target prices and acreage set-asides. Canadian farmers receive the

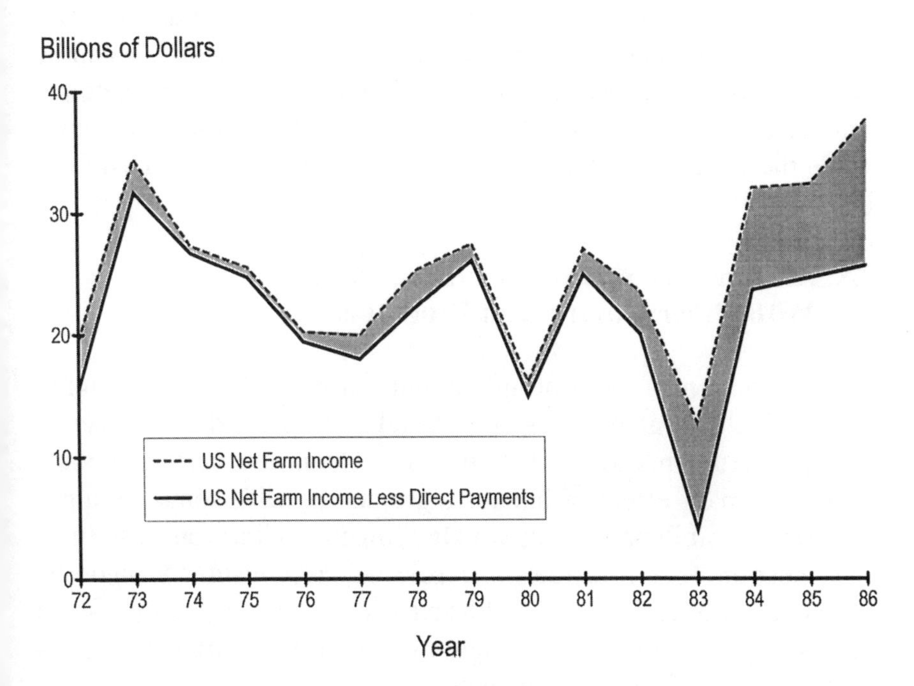

Fig. 3. Direct Government Payments and Realized Net Income, United States, 1972-1986

Source: Agricultural Statistics

market price for grains (at times the U.S. loan rate) but may receive transfers that make up some of the differences between the market price and the target price. In terms of resource use, the two programs have similar effects. Third, from Figs. 1 and 3, it can be seen that the programs have had a different effect on farm income. The Canadian programs in Fig. 1 have tended to stabilize incomes while the U.S. programs in Fig. 3 appear to have destabilized farm income.

7.8 An Analysis of Crop Insurance and Its Interaction With Other Government Programs

The interaction of farm programs with crop insurance is roughly similar in the United States and Canada. While a different set of policy instruments is used in the two countries, most programs have a primary effect of supporting incomes with some having a significant stabilizing effect which competes with crop insurance. This section first illustrates the role of crop insurance and the fundamental problems of moral hazard and adverse selection. Then the effect of other farm programs on crop insurance and its associated problems is investigated.

7.9 The Conceptual Framework

The effects of insurance are represented in Fig. 4. This figure represents either the benefits of insurance on an individual unit of land (for example, an acre or a farm) or insurance benefits for an entire farm economy given the level of plantings. The horizontal axis represents the quantity produced which, on an individual acre of land, would be yield. The vertical axis represents total revenues and costs. Let the market revenue curve be represented by R. At a microlevel or in an open economy where market prices are taken as fixed, it increases linearly with production and is a ray from the origin with slope P_0. This line rotates upward for higher prices and downward for lower prices and is thus represented as conditional

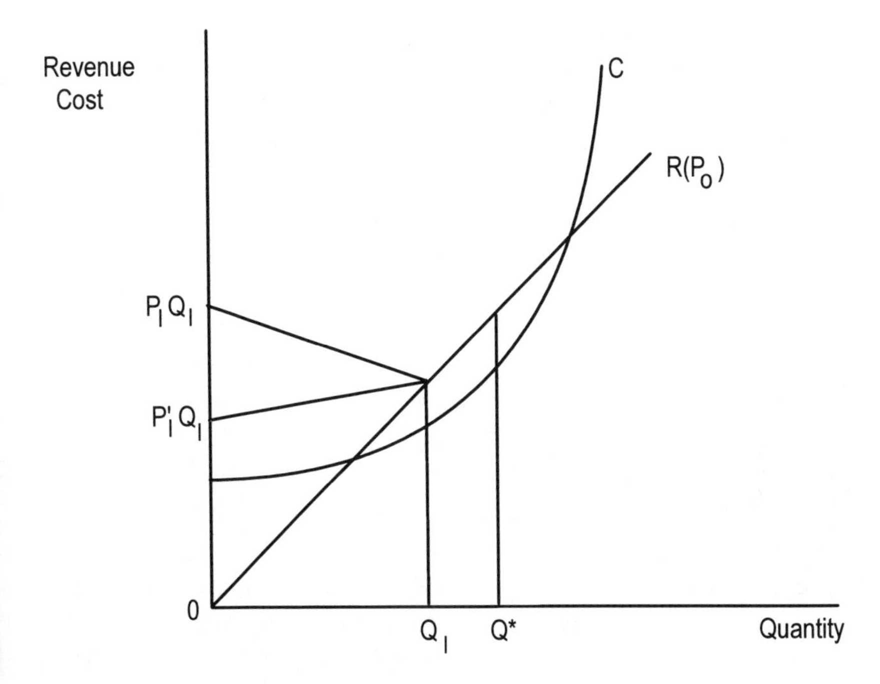

Fig. 4. Effect of crop insurance on farm revenue and profit

on P_0. Costs of planned production on acreage which qualifies for crop insurance are represented by the schedule C. This schedule increases to the right representing the relationship whereby more production is obtained by application of more productive inputs and effort per unit of land. It increases at an increasing rate following the law of diminishing returns. The intercept of the C schedule is positive representing the fact that insurance cannot be obtained on acreage which is not planted at some minimal level. In a free-market case, farm profits are maximized by choosing the point Q* where the positive difference between R and C is maximized (the slope of C is the same as for R). The quantity of inputs and effort per unit of land are then applied accordingly.

Now consider the effect of crop insurance. As federal crop insurance works in both the United States and Canada, the farmer is offered an insurance contract for a specified yield level at a specified price. (While several alternative yield and price levels are offered, the farmer must choose in advance a specific level for both when signing an insurance contract.) An indemnity payment is received only if yield falls below the insured level. The indemnity payment is equal to the insured price level times the quantity by which actual yield falls below the insured yield. The determination of indemnity payments and total revenue is represented in Fig. 5. Fig. 5(a) represents the case where the market price is below the insured price and Fig. 5(b) represents the case where the market price is above the insured price. In each case, the insured yield is represented by Q_I and the insured price is P_I. The actual yield is Q_0 and the market price is P_0. In Fig. 5(a), the insurance indemnity is Q_0abQ_I, market revenue is $0P_0cQ_0$, and total revenue is $0P_0cabQ_I$.

In this case, crop insurance alters the revenue schedule in Fig. 4 as follows. If production is Q_I or more, then revenue is the same as without insurance since no indemnity payment occurs. If production is zero, then revenue is P_IQ_I as shown on the vertical axis of Fig. 4. If the insured price is above the market price, then this point on the vertical axis is above revenue at the insured quantity Q_I. If production is at any level between zero and Q_I, then revenue lies on a straight-line segment connecting the revenues at zero and Q_I such as shown in Fig. 4. Total revenue is $P_0Q_0 + P_I(Q_I - Q_0)$ which is linear in Q_0. For the case where the insured price is below the market price, the analysis is the same except that the revenue at zero quantity in Fig. 6 is below the revenue at the insured quantity such as at P_IQ_I.

To make this analysis operational, two additional considerations are needed. First, in addition to altering the revenue schedule, insurance also alters the cost schedule. Costs are increased at all quantities by the amount of the insurance premium. Thus, the C schedule moves upward in vertically parallel fashion to, say, C′ as illustrated in Fig. 6. Second, crop yield insurance considerations are only meaningful in the context of random production. If cost is incurred on the basis of planned production, then all that is needed is to convert the revenue schedule into an expected revenue schedule. For the free-market case, this is trivial

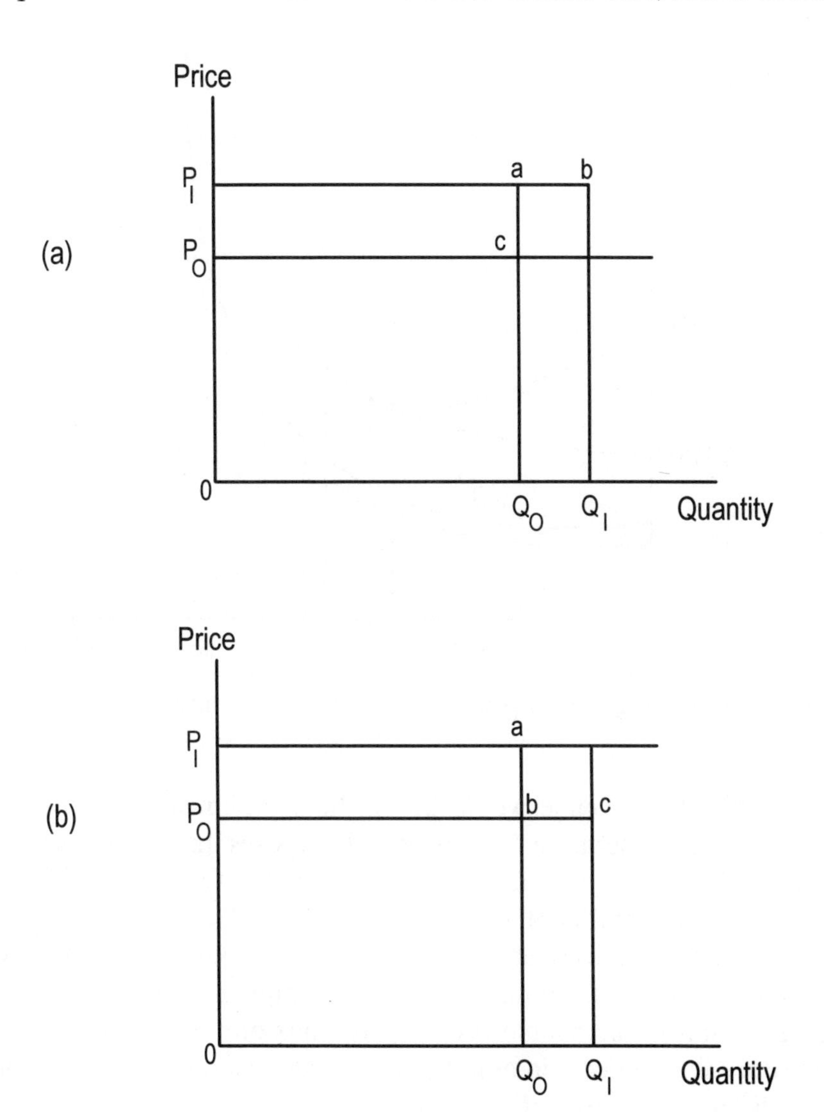

Fig. 5. Effect of crop insurance on farm revenue

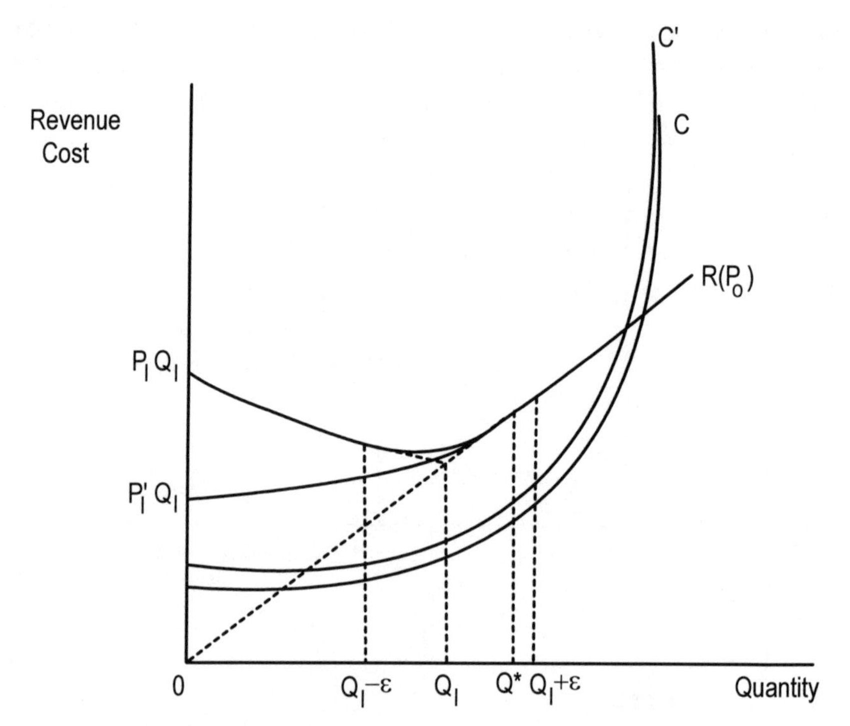

Fig. 6. The moral hazard problem with crop insurance

since expected revenue would lie on the same linear relationship. For other cases, suppose for conceptual purposes that when planned production is Q*, actual production is either $Q^* + \varepsilon$ or $Q^* - \varepsilon$ each with probability 0.5. Then the expected revenue schedule would be the same as above except within a quantity distance e from the kink point. In these regions, computing expected revenue would result in the simple average of points on the actual revenue schedule at distance e in either direction as illustrated in Fig. 6 for two different insured prices, P_I and P'_I.

7.9.1 The Moral Hazard Problem. An important problem related to moral hazard is apparent from Fig. 6. If the insured price is too high, then a serious inducement toward moral hazard results. If the insured price is above the market price, then a farmer is better off letting his planned production fall toward zero since that is where the excess of expected revenue over cost is greatest. On the other hand, if the insured price is sufficiently below the market price, then the quantity at which the excess of expected revenue over cost is maximized may not change significantly as insurance is purchased. The insured price which is sufficiently low enough for this result depends on the elasticity of the cost curve. If the cost curve is highly elastic below Q^*, then the insured price only needs to be slightly below the market price. This point illustrates how moral hazard may be a serious problem for some crops and not for others depending on the relative cost structures.

7.10 The Adverse Selection Problem

Another important problem of insurance that can be illustrated clearly in the framework of Fig. 6 is adverse selection. In Fig. 7, two farmers are considered, one with cost curve C_1, and the other with cost curve C_2. Both are offered the same insurance policy with insured price and quantity P_I and Q_I, respectively. To avoid complicating the figure, suppose the insurance premium is minimal so the same cost curves can be used whether insured or not. Without insurance, Farmer 1 has optimal production Q^*_1 with profit cd. With insurance, Farmer 1 has optimal production Q'_1 with significantly greater profit ab. Farmer 2, however, gains no benefit from insurance and has optimal production Q^*_2, in either case, with profit ef. Furthermore, in this case, the farmer who finds insurance beneficial will alter behavior (moral hazard) so that an insurance indemnity is received in every period.

In the case of Fig. 7, an insurance scheme described by P_I and Q_I, which might be viable if both farmers participated, could

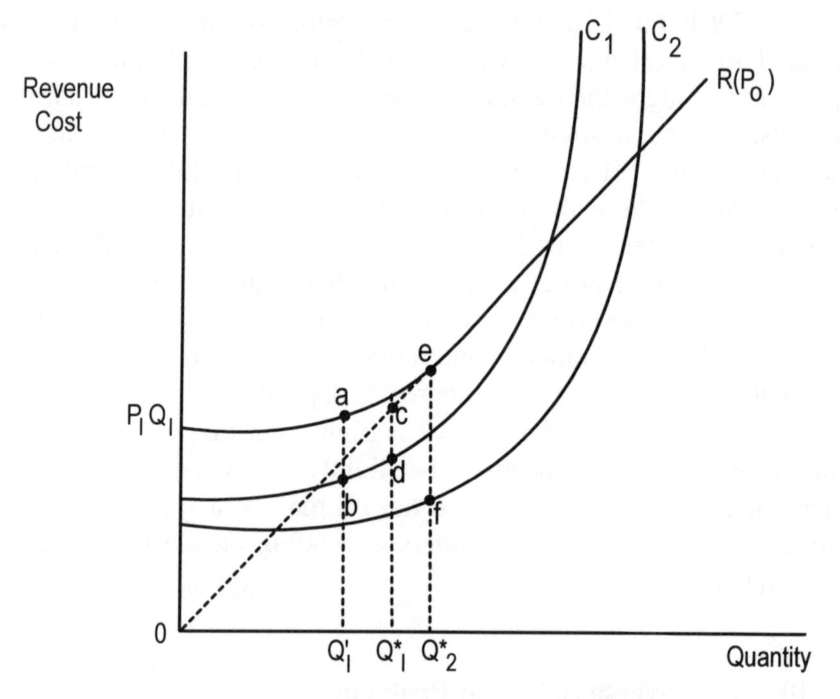

Fig. 7. The adverse selection problem crop insurance

suffer serious losses due to adverse selection because only the farmer more likely to receive an indemnity chooses to participate. The farmers more likely to benefit from insurance are those with lower yields, less elastic cost curves, and possibly greater risk (larger ε causes the smoothed segment of the expected revenue schedule between $Q_I - \varepsilon$ and $Q_I + \varepsilon$ in Fig. 6 to reach farther up the free-market revenue schedule).

7.11 Interaction of Crop Insurance with Price Enhancing Programs

The remainder of this paper examines the problem of how alternative government programs may alter the revenue schedule and thus interact with crop insurance. Most government programs alter the output price received by farmers. This would be true, for example, of the Western Grain Transportation Act in Canada or in the case where the target price is above the market price in the United States. Consider Fig. 8 where initially price is P_0 with the associated expected revenue schedule $R(P_0)$. If farmers have cost schedule C, then crop insurance with insured price P_I and quantity Q_I is doomed to failure because moral hazard will induce farmers to participate in the insurance program and let their production fall toward zero. If, on the other hand, an output price-enhancing program raises the farmers' output price to P1, then the expected revenue schedule shifts to $R(P_1)$ where the optimal output is Q^*_1. The way the cost schedule is drawn in Fig. 8, this could result in participation in the insurance program with little problem of moral hazard. Thus, Fig. 8 demonstrates how output price enhancing programs can interact significantly with federal crop insurance even

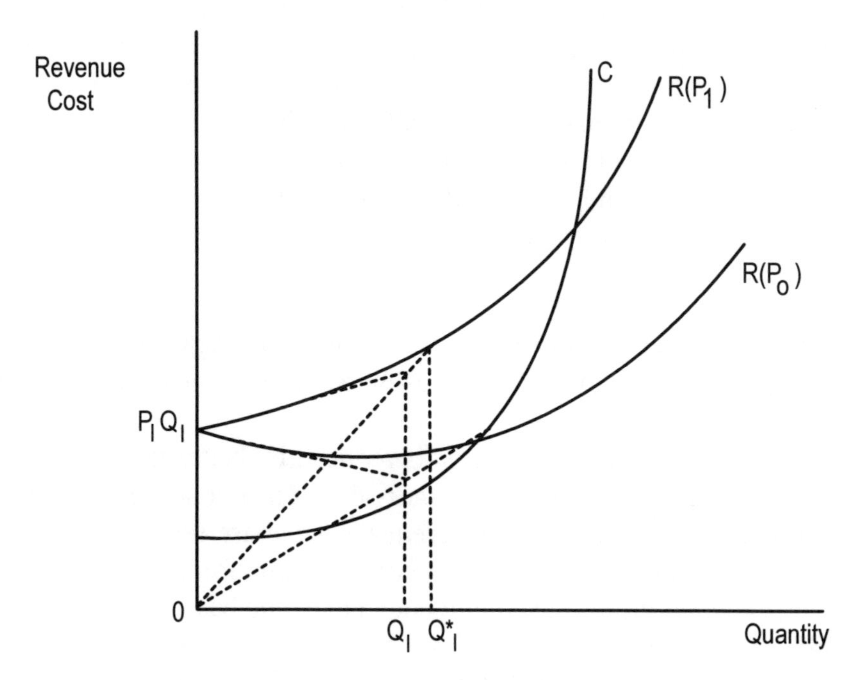

Fig. 8. The effect of price enhancing programs on crop insurance

when the other programs are using controls not specifically designed to influence stabilization. Raising prices received by farmers through other programs without raising insurance prices tends to reduce moral hazard.

7.12 Interaction of Crop Insurance with the Conservation Reserve Program

Several other farm programs can also help to reduce the problem of adverse selection. For example, in the United States, the Conservation Reserve Program is designed to provide an incentive to remove environmentally sensitive lands from production. These lands tend to be of poor quality (higher cost and lower yielding). Thus, the cost of production on these lands likely compares to the cost of producing on other lands much as the cost curve C_1 in Fig. 7 corresponds to the cost curve C_2. If the Conservation Reserve Program is successful in removing such lands from production, then the problem of adverse selection is reduced. In drawing these conclusions, however, it is assumed that such differences in land quality are not already adequately incorporated into the insured yield levels. Since insured yields are established on a farm basis and land quality varies within each farm, this would generally not be the case.

7.13 The Interaction of Crop Insurance with Agricultural Stabilization and Conservation Service Data Generation

The problem of adverse selection is reduced by better tailoring the insured yields and premiums to individual farm conditions. This tailoring, however, requires an adequate data base regarding yield histories. In the United States, farm programs for crops such as wheat and feed grains have deficiency payments based on program yields, the computation and administration of which has led to a substantial data base on farm yields. To administer these programs, the Agricultural Stabilization and Conservation Service generates yield data that is tailored to local conditions. These yield data have been used to tailor FCIC insured yield choices to individual farm conditions where better data are not available. Thus, this data enhancing feature of farm programs has contributed to reducing the problem of adverse selection.

7.14 Interaction of Crop Insurance with Disaster Assistance Possibilities

Next consider the effect of programs such as disaster assistance which have the specific objective of income maintenance in disaster years. These programs, in effect, provide free crop insurance except that ad hoc disaster assistance may be difficult to anticipate. The effects of disaster assistance can thus be reflected in the context of the conceptual framework of this paper in much the same way as those of crop insurance. The major differences are that disaster assistance is usually made available at a lower yield level than crop insurance, and payouts are usually at a level considerably below the market price (so that the initial segment of the revenue curve is positively sloped). A crucial consideration is whether the amount of crop insurance indemnities limits the amount of disaster assistance that can be received or vice versa. In Canada, the two are independent while in the United States the sum of the two that can be received has been limited for some crops in some cases.

Consider Fig. 9 where at price P_o the revenue curve is AE. With cost curve C, the profit maximizing production is Q^*_1. If a disaster assistance program is provided, it places a lower bound on revenue of $P_D Q_D$, with the sum of market revenue and disaster assistance following BE which approaches the free market revenue line AE as production rises from zero. The vertical difference between these two curves represents the expected amount of disaster assistance as a function of the planned production level. The slope of the lower portion of the BE curve depends on what

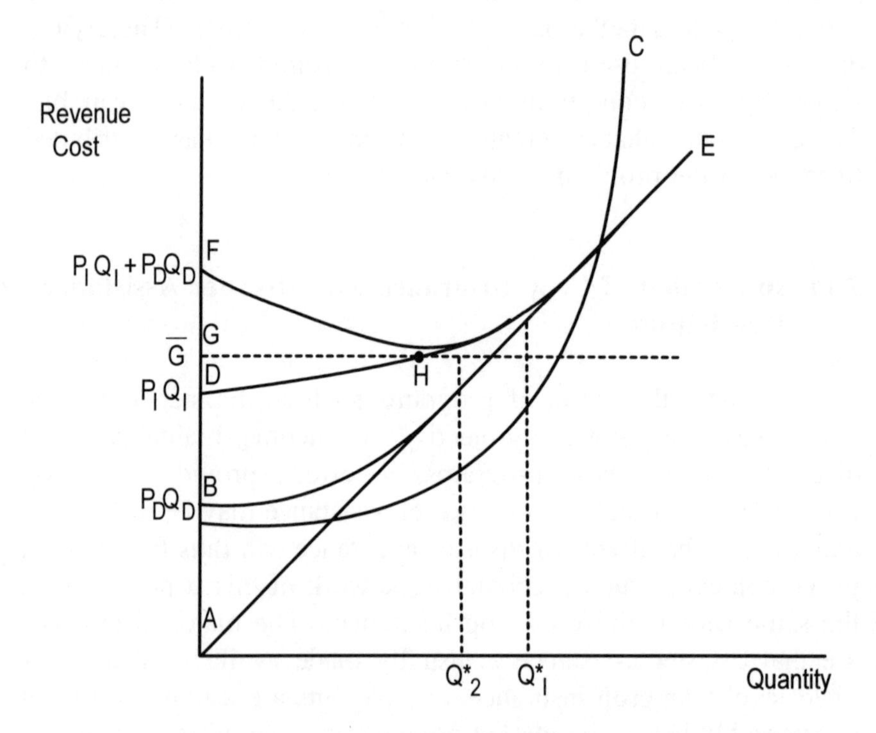

Fig. 9. The effect of disaster assistance potential on crop insurance

percentage of market losses are expected to be made up by disaster assistance and on how much the probability of an ad hoc disaster assistance program increases as production falls toward zero. If these two factors are sufficiently small, then moral hazard is not a major problem (the profit maximizing production remains close to Q^*_1).

Now suppose a crop insurance program is added with insured price P_I and insured yield Q_I in which case the expected revenue curve without disaster assistance follows DE. The amount of the insurance indemnity payment as a function of the planned production level is represented by the vertical difference between the DE and AE curves. As a result of moral hazard, the private profit maximizing production plan falls to Q^*_2.

If, in addition to the crop insurance program, a disaster assistance program is also likely when production is low, and disaster assistance payments are independent of crop insurance indemnity payments, then the expected disaster assistance payments should be added vertically, thus obtaining the FE expected revenue curve. The problem is that the probability of a disaster assistance program now exacerbates the moral hazard associated with crop insurance. Farmers are better off if they let planned production fall toward zero (or as low as can go undetected). Thus, the perception of a positive probability of ad hoc disaster assistance in disaster years on the part of farmers can be an important factor contributing to the failure of a crop insurance program, quite apart from the reduced participation in crop insurance which it induces.

On the other hand, if disaster assistance is not allowed for farmers participating in crop insurance, or a farmer has to choose to receive only one or the other, then the return earned on the crop insurance premium is reduced to the vertical difference between the DE and BE curves. Accordingly, such a limitation will reduce participation in the crop insurance program.

Even if a simple cap is imposed on the sum of crop insurance indemnity payments plus disaster assistance payments, as in the United States, the effect can be to reduce crop insurance participation and/or increase the moral hazard of crop insurance participants. For example, suppose that the sum of payments plus market revenues is limited by G so that expected revenue with crop insurance participation follows GHE. Then, because the cost curve is increasing, expected profits increase as planned production is pushed as far as possible toward zero. Furthermore, the benefits of crop insurance, given the likelihood of disaster assistance, are the vertical difference between GHE and BE which may be less than the vertical difference between DE and AE which applies when disaster assistance is not a possibility. Thus, because the expected benefit of crop insurance may be lower when disaster assistance is possible, participation in crop insurance may be reduced.

7.15 Interaction of Crop Insurance with Income Stabilization Programs

Finally, consider the interaction of crop insurance with programs that are designed to stabilize incomes. The best example is the Western Grain Stabilization Act in Canada. In this case, the program increasingly supplements market revenues as revenues fall toward zero, but payments into the program are made at a constant percentage (4 percent) of revenue up to some maximum ($60,000). Thus, while the lower end of the revenue curve is bent upward much as in the case of crop insurance, the upper end of the revenue curve is rotated downward. For example, in Fig. 10, the free market revenue curve is AE. If farmers participate in the Western Grain Stabilization Act, they contribute a percentage of revenue to the program. This means that the revenue curve rotates clockwise to AC. If point C corresponds to the maximum amount of revenue on which this contribution is made, then the revenue curve for participation above C (CD) is parallel to AE. Finally, as payments to farmers under the program are represented by the vertical difference between BC and AC (higher payments at lower market revenue levels), then the revenue schedule under the program follows BCD.

This will tend to exacerbate the problem of moral hazard, in two ways. First, if the vertical difference between BC and AC is decreasing in production, and the payments from the Western Grain Stabilization Act are independent of crop insurance indemnities, then the problem of moral hazard is exacerbated just as in the case of disaster assistance in Fig. 9. Second, the effect of rotating the revenue schedule downward at higher production levels is much the same as the effect of reducing the output price as demonstrated in Fig. 8. Thus, moral hazard is increased accordingly.

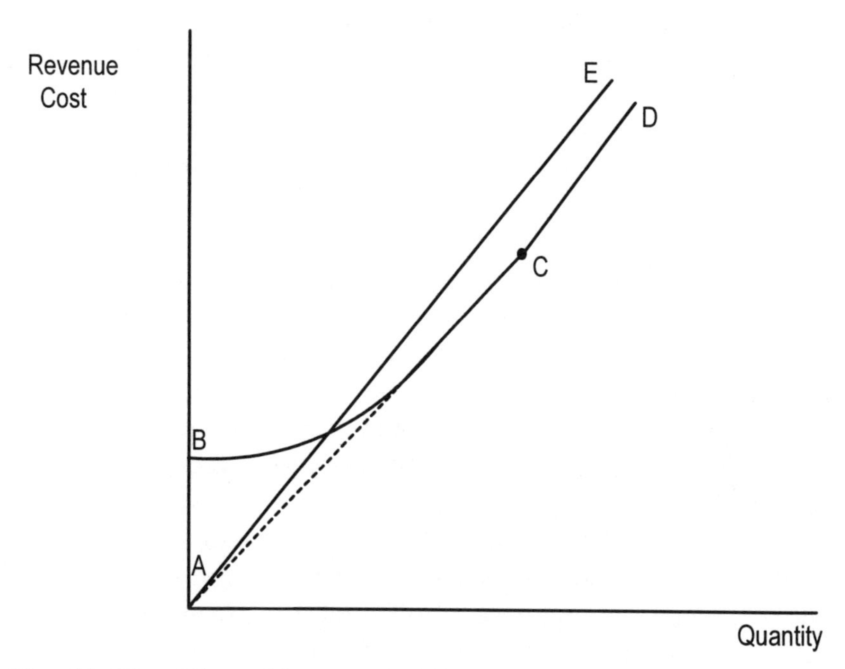

Fig. 10. The effect of income stabilization programs on crop insurance

7.16 Summary and Conclusions

In Canada crop insurance is used by a large percentage of producers. Participation in crop insurance and the size of indemnity payments received is independent of other farm programs. In the United States, participation has been more varied, with high participation in the farming regions similar to the Canadian Prairies (Montana and North Dakota) and low participation elsewhere. In the United States, the crop insurance program has not been independent of other programs. The total amount of disaster assistance payments plus crop insurance indemnity payments is occasionally limited, and this reduces the attractiveness of crop insurance. On the other hand, participation in crop insurance has been augmented occasionally when its purchase has been made a condition for the receipt of disaster assistance and the right to participate in other farm programs.

Generally, crop insurance in both countries has been inadequate to support and stabilize farm incomes as evidenced by the proliferation of other farm programs for this purpose, in particular ad hoc disaster assistance programs. In the Canadian case, one inadequacy of crop insurance from the standpoint of stabilization is its use of the market price, itself uncertain, as a basis for insurance. In the United States, the low level of participation in crop insurance has been a primary factor motivating ad hoc disaster assistance bills.

This paper develops a conceptual framework to examine the interaction of crop insurance with other farm programs. The results show that other farm programs can interact significantly with crop insurance. In particular, raising farm level prices through deficiency payments in the United States or transportation subsidies (the Western Grains Transportation Act) in Canada tends to reduce problems of moral hazard. To the extent that production histories developed under participation in other programs is used to better tailor insured yields to individual farm situations, the problem of adverse selection is also reduced. To the extent that disaster assistance program benefits are additive to crop insurance indemnity payments, they tend to increase the problem of moral hazard. On the other hand, to the extent that disaster assistance program benefits are not additive, they tend to reduce participation in crop insurance, thus reducing the benefits of crop insurance. Finally, income stabilization programs such as the Western Grain Stabilization Act tend to exacerbate the moral hazard problem of crop insurance.

These results may help to explain why crop insurance participation levels have been higher in Canada but they also suggest a greater problem of both moral hazard and adverse selection in Canada than in the United States. Thus, the results also suggest additional reasons why higher rates of subsidy have been required in Canada. In the case of both countries, the results imply that crop insurance policies must be determined in the context of the entire package of farm programs, taking account of the relevant interactions between crop insurance and other programs.

In 1991 the Gross Insurance Program in Canada replaced crop insurance. It has wide coverage in that a farmer can collect from the government if commodity prices are low, even if his yields are high, which would have made him ineligible for a payout under the traditional crop insurance program. The moral hazard and adverse selection problems of the type discussed above are still present under the revenue insurance program and the earlier analyses can easily be extended to apply.

References

Agriculture Canada (1989) Western grain stabilization annual report, 1987-88, Ottawa

Fulton M, Rosaasen KA, Schmitz A (1989) Canadian agricultural policy and prairie agriculture. Economic Council of Canada, Ottawa

Statistics Canada. Agricultural Statistics, Ottawa

United States Department of Agriculture (1987) Agricultural statistics. Economic Research Service, Washington

Part III

Applications and Policy Studies

Chapter 8

An Empirical Analysis of U.S. Participation in Crop Insurance

R.E. JUST[1] and L. CALVIN[2]

8.1 Introduction

While the economic analysis of agricultural crop insurance has attracted increasing attention in the United States over the last decade, most of the work has been theoretical or hypothetical. For example, a number of papers focus conceptually on the problem of why private all-risk crop insurance markets do not emerge (for example, Nelson and Loehman 1987). Several papers investigate the potential explanation associated with adverse selection (Ahsan, Ali, and Kurian 1982; King and Oamek 1983; Skees and Reed 1986), moral hazard (Chambers 1989), and pooling inabilities (Quiggin 1990). Other papers suggest insurance schemes which eliminate or reduce moral hazard by basing indemnities on area yields or weather, limiting them to costs of production, or limiting them according to previous years' claims (Halcrow 1949, Miranda 1990, Lee 1953, Quiggin 1986, Zulauf and Hedges 1988, Lambert 1983, Rogerson 1985). However, very few studies have attempted to examine and explain the low participation levels of U.S. farmers in federal crop insurance schemes that are underwritten by the federal government. Seemingly, the standard problems of moral hazard and adverse selection should only tend to induce insurance participation by those more likely to benefit from insurance. Whether this is actually the case is discussed below.

---

[1]Department of Agricultural and Resource Economics, University of Maryland, College Park, MD 20742, USA.

[2]Economic Research Service, U.S. Department of Agriculture, Washington, DC, USA.

8.1.1 Empirical Analysis of Crop Insurance Participation. Most previous analyses of participation in crop insurance rely on county level data. Gardner and Kramer (1986) used county level data for 1979 to investigate how insurance participation responds to the rate of return from insurance. They estimated that a county's participation rate would increase by about 0.3 percentage points for each 1 percent increase in the rate of return to insurance. Nieuwoudt, Johnson, Womack, and Bullock (1985) used both cross section and time series county level data in a similar investigation of insurance participation rates. Hojjati and Bockstael (1988) used a more sophisticated discrete/continuous (insurance/acreage allocation) choice model to examine insurance participation and land allocation. Unfortunately, the complexity of their model is too great to allow practical application when more than 2 crops are involved, so much of the benefits of the sophistication are likely illusive. They also apply their model using county level data finding results similar to those of Gardner and Kramer. For example, they find that increasing the insured yield level to the 85 percent level would increase participation by 12 percent compared to the current case where maximum coverage is at the 75 percent level.

By relying on county level data, all of these studies miss the within-county heterogeneity among farms that may explain much of the variation in insurance participation. Two counties with the same average yield may have very different participation rates because of a much wider distribution of year-to-year yield in one county than in another.

Before this project began, there were no nationwide empirical analyses of micro level data to determine what actually motivates farmer participation in crop insurance programs in the United States. Skees (1988) reports on a number of crop insurance issues which were investigated with micro level data from Western Kentucky but stops short of developing an empirical model explaining farmers' crop insurance decisions. Patrick (1988) uses hypothetical data on what individual Australian wheat farmers would pay for insurance to analyze the crop insurance decision.

The only study that has attempted to explain U.S. crop insurance decisions econometrically at the micro level is the study by Calvin (1990) using an earlier incomplete version of the data analyzed here. Nevertheless, such an analysis is crucial to ascertain the viability of alternative public or private insurance plans.

8.1.2 The Purpose of This Paper. Crop insurance has become a focal point of interest surrounding the U.S. 1990 farm bill as the debate on whether to continue subsidized federal crop insurance has increased. Several recent reports of the U.S. General Accounting Office have called for studies to increase understanding of farmers' failure to participate in federal crop insurance programs particularly given the high levels of federal subsidies (U.S. General Accounting Office July and August, 1988). In December, 1988, the U.S. Office of Management and Budget requested that the Economic Research Service evaluate federal crop insurance. The University of Maryland-Economic Research Service 1989 crop insurance survey is a part of that endeavor.

In this survey, a Computer Assisted Telephone Interview (CATI) survey, the same farmers as were included in the 1988 Farm Cost and Returns Survey (FCRS) were interviewed. The FCRS is an annual design survey conducted by the National Agricultural Statistical Service for the Economic Research Service. The FCRS survey compiles a wide range of data characterizing the production, input, and cost situation of individual farms. The CATI survey provides data on attitudes, experience, and information surrounding farmers' decisions to participate in federal crop insurance. These data were also enhanced by using actual crop insurance policy information for individuals from the Federal Crop Insurance Corporation (FCIC). Data from all three sources (CATI, FCRS, and FCIC) were merged by farm for the econometric analyses reported in this study. Additional data included FCIC insurance rate data by county and Agricultural Stabilization and Conservation Service (ASCS) data on county yields.

This paper reports on the results of analyzing these data. The next section briefly describes some details and experience regarding the multiple peril crop insurance program in the United States. In the following section a number of alternative explanations are offered for the low participation in federal crop insurance in spite of heavy government subsidies. With this background, the paper then turns to the development of a framework in which to analyze the data. Finally, the paper gives a detailed description of the survey which was conducted and a report of the analysis and the associated explanation of low participation.

8.2 Federal Crop Insurance in the United States

Federal crop insurance in the United States was introduced with the Federal Crop Insurance Act of 1938. Under this legislation, the Federal Crop Insurance Corporation (FCIC) was created to insure farmers against losses from all forms of yield risk such as drought, hail, wind, flood, frost, disease, and insects, without regard to cause. However, federal crop insurance was not an important part of the U.S. farm program until the Federal Crop Insurance Act of 1980. Previously, participation rates were low and, at times, few crops and limited numbers of counties were included in the program. With the 1980 Act, coverage of both crops and counties was increased and a 30 percent subsidy of premiums was undertaken by the federal government. Unfortunately, participation rates remained below 25 percent until 1989 when participation was required as a condition for receiving drought assistance in 1988.

8.2.1 How Federal Crop Insurance Works in the United States. For most crops under this program, farmers who purchase insurance choose one of three yield levels and one of three price levels at which to insure. The alternative yield levels are 50, 65, and 75 percent of the farmer's Approved Production History (APH) yield. The APH yield is calculated as a ten year average of yields obtained on the farm if actual verified yields are available. If a sufficient verified yield history does not exist, then the individual's ASCS yield is used in place of the APH yield. If an individual ASCS yield is not available, then the FCIC assigns a yield to the farmer. Generally, the FCIC estimates expected market price and sets the highest price guarantee at approximately that level, but in no case less than 90 percent of that expected price. For example, in 1989 the alternative insured price levels for corn were $1.50, $2.00, and $2.60 per bushel. For soybeans in 1989, however, the high insured price was based on 85 percent of the average futures price over the last week of March for the November contract.

The premium paid by farmers depends on the selected yield and price levels, the APH yield, and the county level premium rate which reflects local risk conditions. The federal government subsidizes the premium paid by farmers as follows. If the farmer insures at the 50 or 65 percent levels, then the government pays 30 percent of the farmers premium. If the farmer insures at the 75 percent level, then the government subsidizes the premium at a rate equal to 30 percent of the premium for the 65 percent level of insurance.

Farmers receive an insurance indemnity payment if the average yield on their entire farm falls below the insured yield level. The indemnity payment is equal to the insured yield less the actual yield, evaluated at the insured price. For example, in Fig. 1, if the insured yield is chosen at the 65 percent level and the actual yield is below that level, then the per acre indemnity payment is represented by the cross hatched area, dcy_1y_0. This income is in addition to the market revenue per acre of p_0ay_00. Total revenue per acre may be smaller than normal because either (1) the insured percentage of APH yield is small, or (2) the APH yield is smaller than the farmer's true normal yield, or (3) the insured price is below the normal market price.

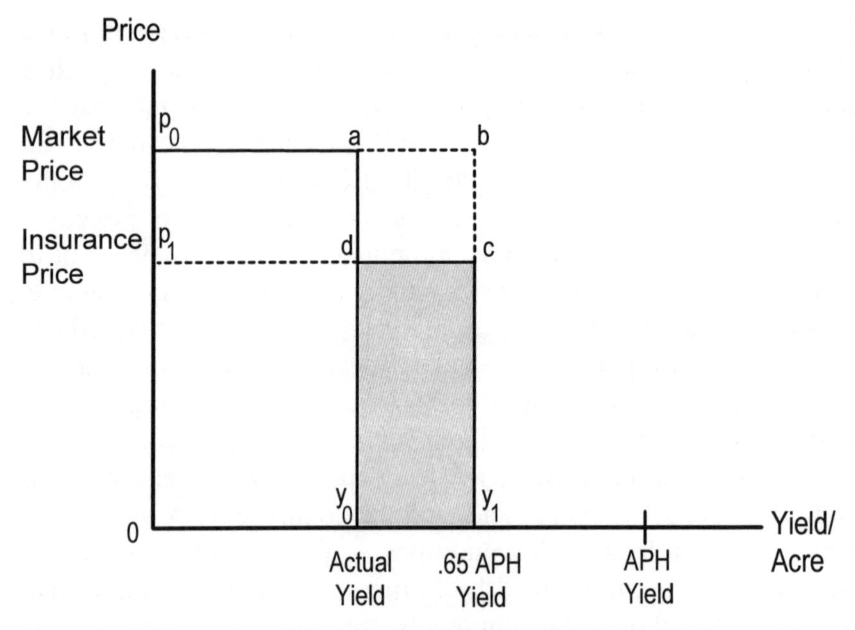

Fig. 1. Indemnity payment under FCIC multiple peril crop insurance

8.2.2 Failure of the Federal Crop Insurance Act of 1980. With the 1980 Act, the delivery of federal crop insurance was shifted to the private sector. The FCIC uses two basic delivery systems. With the first, which handled about 85 percent of insurance sales in 1988, private insurance companies handle all aspects of sales, service, and loss adjustment on their own with the FCIC reinsuring these companies against extraordinary losses (in excess of 15.375 percent of premiums in 1989). In the last few years, this has resulted in FCIC compensation to reinsured companies equal to about one-third of collected premiums. With the second system, master marketers work as contractors with the FCIC to sell and service policies written by FCIC with the FCIC overseeing loss adjustments. In 1989, master marketers were compensated at the rate of 20 percent of collected premiums. The insurance policies are identical in the two cases.

Although these changes brought about a major increase in participation from 9.6 percent in 1980 to 24.5 percent in 1988 (as a percentage of planted acres of insurable crops), this increase was far short of the program goal of 50 percent. (A higher participation rate was reached in 1989 but only by requiring purchase of crop insurance by those receiving disaster assistance in 1988.) This disappointing participation level occurred in spite of a heavy increase in government outlays for crop insurance which exceeded $4.2 billion between 1980 and 1988 with a loss ratio (indemnities divided by premiums) averaging 2.05 (U.S. General Accounting Office 1989). (A loss ratio less than about .95 is generally regarded as necessary for private insurance viability.) This high loss ratio is particularly disturbing given the program goal of gradually privatizing the crop insurance program and eliminating government subsidization. While the delivery of federal crop insurance has been largely privatized, government subsidization has increased.

8.3 Potential Explanations of Failure of Federal Crop Insurance

A variety of reasons potentially explain the low federal crop insurance participation rates in spite of heavy government subsidization. These reasons include heterogeneity of risk, heterogeneity of average yields, other aspects of heterogeneity, the associated problems of adverse selection and moral hazard, and the expectation of free insurance through ad hoc disaster assistance programs. Not all farmers need crop insurance and the benefits to participating are not uniform across all farmers.

8.3.1 Heterogeneity of Risk. One reason is that only high risk farmers can take advantage of the program. For example, if a farmer's yields never fall below 75 percent of normal (and the farmer's APH yield is the true normal yield) then the farmer will never be able to collect an indemnity payment. In contrast, if a farmer's yield is either 50 percent or 150 percent of normal with probability .5 each, then an indemnity payment will be collected half of the time. Thus, two farmers with the same normal yield may receive quite different benefits from participation. This problem of heterogeneity of risk explains how the federal crop insurance program can be drawing limited participation in spite of high subsidies.

This problem is represented in Fig. 2. Panels (a), (b), and (c) represent three farmers with different risk levels but the same normal yield level, μ. In each case, the curve describes the relative frequency with which each yield level occurs. In the case of farmer (a), the risk is low enough that an indemnity payment can never be collected because yield will never be below 75 percent of normal. With farmer (b), the risk is low enough that an indemnity payment cannot be collected if the farmer is insured at the 50 percent level. If the farmer is insured at the 75 percent level, the probability of collecting an indemnity payment is equal to the shaded area in panel (b). However, the expected amount of the indemnity payment will be small if most of the probability density below $.75\mu$ is near $.75\mu$ (because the indemnity payment is based on the difference between $.75\mu$ and the actual yield). The case in panel (c) represents a high risk farmer. Here the probability of collecting an indemnity payment if the farmer is insured at the75 percent level is represented by the shaded area. In this case, however, much of the probability is far below the $.75\mu$ level so the expected size of the indemnity payment is substantial.

Casual analysis of the data suggests the importance of this explanation for low insurance participation. In many areas, the probability of at least a 25 percent yield loss is low. However, in the case of dry land wheat farming in Montana and North Dakota, the probability of large yield losses is high, with crop failures occurring in some areas in one out of every three or four years. This seemingly explains the high participation rates in crop insurance for wheat in these two states compared to other states (Table 1).

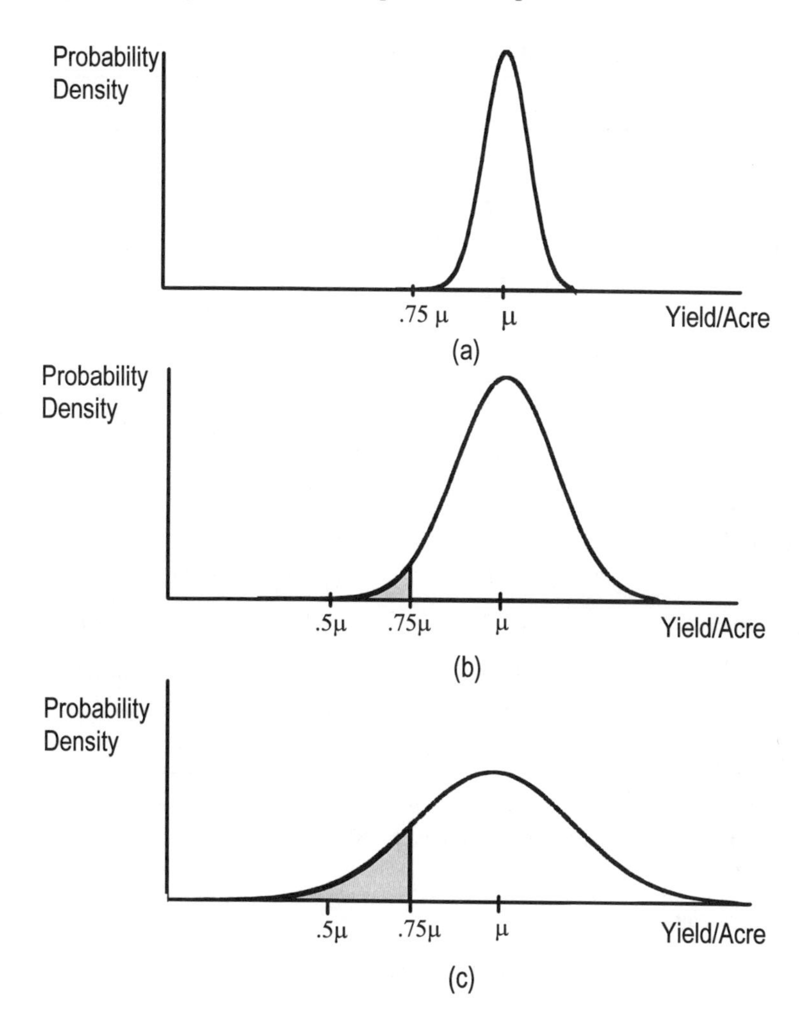

Fig. 2. Effect of risk on ability to collect indemnities

8.3.2 Heterogeneity of Average Yields. Another reason for low participation in federal crop insurance programs is that farmers' average yields may not be well reflected in the FCIC yields used for insurance purposes. For example, when a sufficient verified production history does not exist on a farm, the

Table 1. Federal crop insurance participation rates by state, 1987[a]

State	Wheat	Corn	Soybeans
Colorado	0.160	0.218	[b]
Illinois	0.038	0.131	0.102
Indiana	0.023	0.127	0.084
Iowa	0.023	0.330	0.295
Kansas	0.156	0.233	0.231
Minnesota	0.373	0.210	0.378
Missouri	0.062	0.311	0.202
Montana	0.706	0.092	[b]
Nebraska	0.302	0.307	0.248
North Dakota	0.511	0.148	0.314
Ohio	0.019	0.081	0.053
Oklahoma	0.093	0.151	0.318
South Dakota	0.170	0.109	0.254
Texas	0.060	0.217	0.444
Washington	0.341	0.001	[b]

[a]The participation rate is the sum of net acres insured divided by the sum of potential insurable acres at the state level. Net acres are acres multiplied by the insured share.

[b]Not applicable.

Source: Federal Crop Insurance Corporation

individual's ASCS yields are used as a basis for establishing insured yields, premiums, and indemnities. Nevertheless, average yields may vary considerably among farmers in the same county because of land quality, managerial ability, production practices, and so on.

Suppose half of the farmers in a county farm on high quality soil and the other half farm on low quality soil. Suppose further that both have the same yield $\bar{\mu}$, for insurance purposes, that is an average across all farmers. Then, as in Fig. 3(a), the

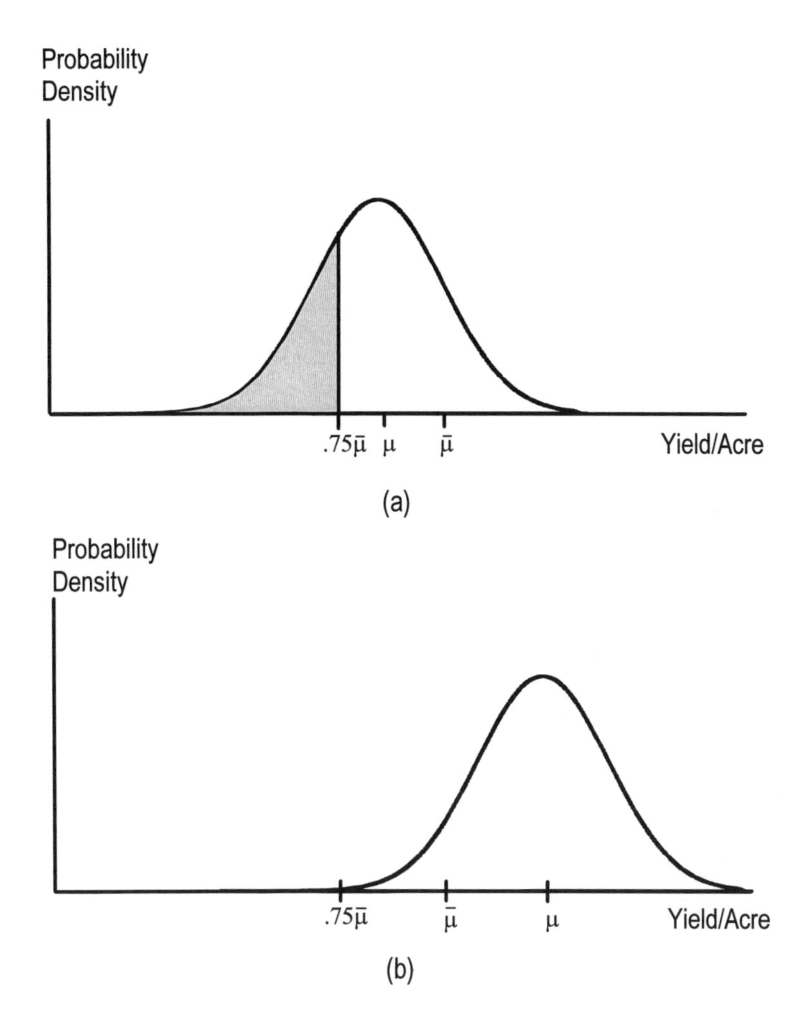

Fig. 3. Effect of the relationship of expected yield and insurance yield on the ability to collect indemnities

farmers on low quality soil will have an actual average yield μ below the insurance yield whereas in Fig. 3(b) the farmers on high quality soil will have an actual average yield μ above the insurance yield. Suppose both have the same level of risk (the probability densities differ only by location). Then the farmers in panel (a) will have a higher probability of collecting an indemnity payment. With sufficient heterogeneity among farmers, the farmers with low quality land may have a substantial probability of collecting indemnity payments while the farmers with high quality land may have no chance of collecting indemnity payments. Again, participation may be low (for example, if few farmers have low quality land) but indemnities may be high among those who participate.

In reality, beginning in 1982, FCIC provisions have allowed farmers to tailor their insurance yields to their own circumstances by providing for APH yields based on verified production history on the individual farm. However, while farmers on high quality land have an incentive to establish a verified production history, a farmer on low quality land has no incentive to keep a verified production history and is better off using the county ASCS yield. Thus, the resulting inequity is similar in spirit to the problem depicted in Fig. 3(a). The applicability of this explanation is difficult to investigate with readily available data (for example, county level data) because it depends on heterogeneity among individual farms. Thus, the analysis of farm-level data is necessary.

8.3.3 Other Aspects of Heterogeneity. Many other aspects of heterogeneity among farms play an important role in the decision to participate in crop insurance. One is farm size. On a large farm with many tracts, a large yield loss is less likely because many causes of yield loss are highly localized (for example, hail). The reception of an indemnity payment depends on whether the average yield across the whole farm unit (for insurance purposes) is less than the elected insurance yield. A farmer may sustain a total loss on one tract and not collect an indemnity because the yield on that tract is averaged together with other tracts. On a farm with only one tract, the probability of, say, a 25 percent yield loss may be considerably higher which, ceteris paribus, means that the smaller (or more localized) farmer gets a higher expected indemnity per acre. Farmers with multiple farms (as defined by farm identification numbers) have the option of calculating losses by farm but are charged an additional 10 percent in premiums for this option.

Another consideration is that farms with higher average yields do not necessarily have proportionally higher variability. The FCIC attempts to take some account of this variation by setting premium rates at a lower level where the area yields are higher. However, in a sample of Western Kentucky farmers, Skees and Reed (1986) found that this adjustment, in which the FCIC assumes constant relative risk among farms, is insufficient and favors farmers with lower average yields.

Irrigation is another important source of heterogeneity. The average yield may be far higher and the relative risk far less on irrigated acreage than on dry land acreage. While some of this difference may be represented in different premium rates on irrigated land, the yields on which the premiums are based may not reflect irrigation differences. For example, in Maryland the ASCS yields do not differentiate by irrigation. Thus, farmers who irrigate part of their acreage are at a disadvantage in collecting indemnities compared to farmers who irrigate none of their acreage (Cawley 1989).

8.3.4 Adverse Selection. All of these aspects of heterogeneity have substantive implications for adverse selection. The problem of adverse selection occurs when agents (farmers) differ in characteristics that affect the probability and size of indemnity payments but these characteristics are not reflected appropriately in the premium structure. With a heterogeneous farm population, insurers have difficulty assessing the risk of individual farmers. Thus, the agents who are more likely to collect indemnities or who collect larger indemnities per unit of premium are more likely to participate. As a result, the insurer must either set the premium structure to reflect the higher risk associated with the individuals who choose to participate (thus offering less fair insurance to others) or a loss will be incurred by the insurer (for example, if the premiums reflect the average characteristics of all agents).

The latter situation apparently prevails in the case of U.S. federal crop insurance (see the evidence presented below). That is, if the FCIC attempts to set insurance premiums at levels that reflect average conditions in U.S. agriculture, then it is likely to incur losses because of adverse selection. The problem is one of asymmetric information. Farmers know more about their likelihood of obtaining alternative yield levels than do insurers formulating crop insurance premiums. To reduce this problem, the FCIC would have to take into account more of the factors that explain heterogeneity among farms (for example, risk, land quality, managerial ability, and farm size) with premium structures reflecting these factors in an actuarially fair way. In assessing these possibilities, it is important to know how much each factor contributes to adverse selection and insurer losses.

8.3.5 Moral Hazard. A related problem of asymmetric information is moral hazard. The problem of moral hazard occurs when an insured farmer may follow less costly or less time consuming production practices when insured, thus increasing the probability or size of indemnity payment. The effect of moral hazard is much like that of adverse selection. That is, the insured agents are more likely to incur the insured event than are the noninsured. Thus, either the insurance premiums must be adjusted to reflect the higher and more likely payments of indemnities to insured agents, or the insurer will incur losses. In this case, however, the higher premiums or insurer losses occur because of endogenous changes in behavior when farmers take out insurance rather than because of inherent differences among farmers.

The extent to which moral hazard explains FCIC losses has not been clear. A definitive determination of moral hazard would require monitoring of production practices and a detection of a change in those practices as a result of the insurance decision. Such a determination, in the case of subtle changes in production practices, is beyond the scope of this paper and apparently beyond the realm of possibility in FCIC practices. However, several indirect approaches can be taken to determine the order of magnitude of impacts of moral hazard. Some of these approaches, which are not feasible with most other data sets, are reported below.

8.3.6 Ad Hoc Disaster Assistance Bills. Another important reason why farmers may not elect to purchase federal crop insurance is that they perceive themselves as receiving some free crop insurance against major losses. Part of the problem is that farmers became accustomed to receiving free federal crop insurance under the disaster assistance provisions of the 1973 and 1977 farm bills. Under these programs, farmers participating in farm programs received payments as a percentage of target prices to cover losses due to natural causes which either prevented crop planting or resulted in low yields.

The enhancement of federal crop insurance under the 1980 Act was intended to replace the federal disaster assistance programs of the 1973 and 1977 farm bills. However, the free-crop-insurance-against-major-disasters mentality has been reinforced by Congress through provision of ad hoc disaster assistance in four out of nine years since the 1980 Act was passed (1983, 1986, 1988, and 1989). Furthermore, some provisions that limit the sum of crop insurance indemnity and disaster assistance payments per acre that can be received have been incorporated into some of these disaster assistance bills. Even though this limitation is rarely effective, it implies that the farmer buying insurance, in effect, has a smaller expected indemnity when Congress is expected to pass ad hoc disaster assistance bills in the event of disaster.

The average magnitude of disaster assistance has been larger in the 1980s than FCIC indemnity payments. Disaster payments under the Disaster Assistance Act of 1988 alone amounted to about $4 billion compared to indemnity payments by the FCIC of only about $1 billion for 1988. While the availability of these payments cannot be anticipated perfectly, farmers have been given a clear signal that the federal government stands ready to bail them out in the case of major and widespread disaster. As a result, a high correlation of yields among producers can make ad hoc disaster assistance an effective substitute for crop insurance. One aim of this study is to determine how experience with disaster assistance programs affects the insurance participation decision.

8.4 Considerations in Data Collection and Model Design

The potential alternative explanations for low insurance participation dictate several types of data that must be included in the analysis. In turn, the model must then be designed to accommodate evaluation of the roles of the associated variables. Some of these considerations in data collection and model design are as follows.

1. First, individualized yield histories or subjective opinions are crucial for adequate assessment of farm-level heterogeneity of both average yields and yield variability among farms and the extent to which heterogeneity of various forms explains the variation in insurance participation. Since yield histories are not uniformly kept across all farms, the necessity of comparable data dictates collection and use of subjective assessments by farmers.

2. Second, what farmers perceive as the probabilities of obtaining less than 50, 65, and 75 percent of their FCIC yield is crucial to the insurance decision and could differ substantially among farmers even if moments such as the mean or variance of yields do not. Collection of such information and measuring its relationship to the insurance decision is necessary in identifying moral hazard and adverse selection, and is crucial in the determination of the probability of an indemnity payment at the farm level. These considerations dictate the need for the CATI survey of individual farmers.

3. Third, farm-level diversification data are crucial in determining the need for insurance. Less specialized farmers are

relatively less affected by a disaster with respect to a single crop and, thus, are in a better position to self insure. Calvin (1990) has demonstrated the importance of diversification in a preliminary study with the FCRS data. Farm-level scale and diversification data such as are included in the FCRS are necessary for the analysis.

4. Fourth, farm-level premium data and the choice of insurance level (both price and yield) are needed to evaluate the insurance decision appropriately. Given the range of insurance choices and associated premiums available to the farmer, a simple indicator of insurance participation is inadequate. Whether a farmer chooses to insure will depend on the premium level and whether insurance is purchased at the 50, 65, or 75 percent level. The choice of insured price will depend crucially on how the premium differs among the various choices. For this reason, the data on actual FCIC policies are useful.

5. The consideration of the demand for crop insurance needs to be crop specific. The insured price level and risk differs among crops. Farmers can choose to insure one crop and not another. Since the sophistication of a model that considers the joint decision of insurance for a group of crops and/or the associated acreage allocation decisions is too complex for tractability, the model structure and analysis are necessarily crop specific.

6. Variables such as wealth, age, education, and the debt-asset ratio can affect attitudes toward risk. Risk attitudes are basic in the insurance participation decision and need to be included in the analysis. Some of these variables are available in the FCRS but some require additional data collection in the CATI.

8.5 A Simple Framework for Analyzing Crop Insurance Participation

In this section, a simple framework is developed in which to evaluate the role of the various factors discussed above in explaining the low participation rates in U.S. federal crop insurance despite heavy government subsidization. Following the basic principles outlined by Just (1990) and by Rausser and Just (1981), the model is developed in a highly structured fashion that constrains the various potential explanatory forces to obey plausible relationships in any estimated model. This tends to prevent peculiar results due to multicollinearity and casual correlation, and increases economic intuition forthcoming from the empirical results.

For an underlying framework with which to examine the FCRS, FCIC, and CATI data, consider the following. Let quasi-rent per acre without insurance be represented by

$$\pi_i = py_i - c_i,$$

where

π_i = quasi-rent per acre from production of the relevant crop on farm i,

p = output price of the relevant crop,

y_i = average yield per acre of the relevant crop on farm i in a given year, and

c_i = variable costs per acre for the relevant crop on farm i.

The output price is considered known at the time of making the insurance participation decision. While prices are not usually known in advance with perfect certainty, at least for the entire amount of the crop, mechanisms such as futures markets, forward markets, and government target and support prices exist to reduce price risk at the time of the insurance participation decision to relatively small levels compared to yield risk. This is true even though prices may be quite unstable from year to year. The considerable clarity that can be gained in conceptual modeling by eliminating complications associated with price risk (and particularly price and yield correlation) appears warranted.

Let each farmer perceive a stochastic distribution for the farm's average annual yield characterized by:

μ_i = the farmer's subjective mean yield per acre on farm i;

and

σ_i = the farmer's subjective variance of yield per acre on farm i.

Suppose insurance contracts offered to the farmer are characterized by four parameters:

$\bar{\mu}_i$ = the FCIC average yield ascribed to farm i;

β = an insurance coverage parameter defining the yield guarantee such that the farmer is insured against yield shortfalls below $y_i = \beta\bar{\mu}_i$;

p_α = an insurance price guarantee level such that the farmer receives an indemnity payment equal to $(\beta\bar{\mu}_i - y_i)p_\alpha$ if the insured event occurs; and

$\gamma_{\alpha\beta}$ = the per acre insurance premium associated with the levels of α and β chosen and the individual's $\bar{\mu}_i$.

Then the farmer's per acre indemnity income is represented by

$$s_i = (\beta\bar{\mu}_i - y_i)p_\alpha\delta \text{ where}$$

$$\delta = \begin{cases} 0 \text{ if } \beta\bar{\mu}_i - y_i \leq 0 \\ 1 \text{ if } \beta\bar{\mu}_i - y_i > 0. \end{cases}$$

Quasi-rent per acre with insurance is thus

$$\pi_i = py_i + (\beta\bar{\mu}_i - y_i)p_\alpha\delta - c_i - \gamma_{\alpha\beta}.$$

The expected size of the indemnity payment is

$$E(s_i) = E[(\beta\bar{\mu}_i - y_i)p_\alpha\delta] = E[(\beta\bar{\mu}_i - y_i)p_\alpha | y_i < \beta\bar{\mu}_i] \cdot P(y_i < \beta\bar{\mu}_i).$$

This is the amount that must be compared to the insurance premium to determine whether participation in crop insurance is profitable on average. For risk averse farmers, it is also important to realize the extent to which this comparison depends on the variability of the farmer's yields.

For purposes of representing the insurance decision, define the following notation:

$$\psi_i^\beta = P(y_i < \beta\bar{\mu}_i) = P(\delta = 1);$$

$$\mu_i^\beta = E[y_i | y_i < \beta\bar{\mu}_i]; \text{ and}$$

$$\sigma_i^\beta = E[y_i^2 | y_i < \beta\bar{\mu}_i].$$

These are the parameters of the truncated distribution of yields that pertains to the insured event. In this case, expected quasi-rent per acre without insurance is

$$\bar{\pi}_i = E(\pi_i) = p\mu_i - c_i,$$

and expected quasi-rent with insurance is

$$\bar{\pi}^*_i = E(\pi^*_i) = p\mu_i - c_i + (\beta\bar{\mu}_i - \mu_i^\beta)p_\alpha\psi_i^\beta - \gamma_{\alpha\beta},$$

with insurance at levels described by α and β. The variance of quasi-rent per acre without insurance is

$$\sigma_\pi = V(\pi_i) = p^2\sigma_i = p^2E(y_i^2) - p^2\mu_i^2.$$

To determine the variance of quasi-rent per acre with insurance, note that

$$\sigma_\pi^* = V(\pi_i^*) = V[py_i + (\beta\bar{\mu}_i - y_i)p_\alpha\delta]$$
$$= V(py_i) + V[(\beta\bar{\mu}_i - y_i)p_\alpha\delta] + 2\,Cov[py_i,(\beta\bar{\mu}_i - y_i)p_\alpha\delta]$$

and that

$$V[(\beta\bar{\mu}_i - y_i)\delta] = E[(\beta\bar{\mu}_i - y_i)^2\delta^2] - [(\beta\bar{\mu}_i - \mu^\beta_i)\psi^\beta_i]^2$$

$$= 0 + E[(\beta\bar{\mu}_i - y_i)^2|y_i < \beta\bar{\mu}_i]\psi_i^\beta - [(\beta\bar{\mu}_i - \mu^\beta_i)\psi^\beta_i]^2$$

$$= \beta^2\bar{\mu}^2_i\psi^\beta_i - 2\beta\bar{\mu}_i\mu^\beta_i\psi^\beta_i + \sigma^\beta_i\psi^\beta_i - [(\beta\bar{\mu}_i - \mu^\beta_i)\psi^\beta_i]^2$$

$$Cov[y_i,(\beta\bar{\mu}_i - y_i)\delta] = E[(\beta\bar{\mu}_i - y_i)\delta y_i] - (\beta\bar{\mu}_i - \mu^\beta_i)\psi^\beta_i\mu_i$$

$$= 0 + E[y_i\beta\bar{\mu}_i|y_i < \beta\bar{\mu}_i]\psi^\beta_i - E[y^2_i|y_i < \beta\bar{\mu}_i]\psi^\beta_i$$

$$- (\beta\bar{\mu}_i - \mu^\beta_i)\psi^\beta_i\mu_i$$

$$= \beta\bar{\mu}_i\mu^\beta_i\psi^\beta_i - \sigma^\beta_i\psi^\beta_i - (\beta\bar{\mu}_i - \mu^\beta_i)\psi^\beta_i\mu_i.$$

Thus, the variance of quasi-rent per acre with insurance is

$$\sigma^*_\pi = V(\pi^*_i) = p^2\,\sigma_i + p^2_\alpha[(\beta^2\bar{\mu}^2_i - 2\beta\bar{\mu}_i\mu^\beta_i + \sigma^\beta_i)\sigma^\beta_i$$

$$- (\beta\bar{\mu}_i - \mu^\beta_i)^2(\psi^\beta_i)^2]$$

$$+ 2pp_\alpha[\beta\bar{\mu}_i\mu^\beta_i - \sigma^\beta_i - (\beta\bar{\mu}_i - \mu^\beta_i)\mu_i]\psi^\beta_i.$$

Now suppose each farmer i decides to insure or not by maximizing a mean-variance expected utility function with constant absolute risk aversion level ϕ_i, that is, crop insurance is (not) purchased if[3]

$$E[U_i(A_i\pi^*_i)] = A_i\bar{\pi}^*_i - \phi_i A^2_i \sigma^*_\pi/2 > (<) E[U_i(A_i\pi_i)]$$

$$= A_i\bar{\pi}_i - \phi_i A^2_i \sigma_\pi/2$$

where

U_i = utility function of farmer i, and
A_i = acreage devoted to the relevant crop by farmer i.

More specifically, insurance is chosen if the additional expected utility per acre with insurance is positive,

$$\Delta EU_i/A_i = \bar{\pi}^*_i - \bar{\pi}_i - (\sigma^*_\pi - \sigma_\pi)\phi_i A_i/2 > 0,$$

that is, if

$$(1) \qquad \Delta EU_i/A_i = (\beta\bar{\mu}_i - \mu^\beta_i)p_\alpha\psi_i^\beta - \gamma_{\alpha\beta} - A_i\phi_i p p_\alpha[\beta\bar{\mu}_i\mu^\beta_i - \sigma^\beta_i$$

$$- (\beta\bar{\mu}_i - \mu^\beta_i)\mu_i]\psi^\beta_i$$

$$- (A_i\phi_i/2)p^2_\alpha[(\beta^2\bar{\mu}^2_i - 2\beta\bar{\mu}_i\mu^\beta_i + \sigma^\beta_i)\psi^\beta_i$$

$$- (\beta\bar{\mu}_i - \mu^\beta_i)^2(\psi^\beta_i)^2] > 0$$

---

[3]Here the acreage is regarded as predetermined and the income from other crops and activities is ignored under the assumption that quasi-rent for the relevant crop is stochastically uncorrelated. This assumption is sufficient with constant absolute risk aversion. These assumptions are difficult to relax in a formal framework without losing tractability. The assumption regarding correlation is necessary because collecting data on correlation was infeasible in the CATI survey for budgetary and federal clearance reasons. The assumption regarding predetermination of acreage is necessary to avoid the sophisticated approach of Hojjati and Bockstael (1988) which becomes intractable if many crops. Although these assumptions are restrictive, the resulting model is believed to capture the important elements of the crop insurance decision better than an ad hoc specification.

for some α and β in the crop insurance choice set.

To empiricize the model, note that the β choice set is .5, .65, and .75, p_α is one of the price insurance levels under the federal crop insurance program for the relevant crop, and $\gamma_{\alpha\beta}$ is the associated insurance premium. Values for ψ^β_i are obtained by means of the CATI survey. Values for μ^β_i and σ^β_i can be inferred by imposing a parametric form on the distribution of yields from values of ψ^β_i for the various values of β, for example, assuming normality. The value of ϕ_i can be represented implicitly as a function of socioeconomic factors and financial status data such as debt, equity, total farm income and risk, off-farm income and wealth situations.

The problem of adverse selection is captured in several ways. First, the heterogeneity of average yields is reflected in the difference between the farmer's actual expected yield and the FCIC's mean yield (for insurance purposes). This difference reflects how the actual insurance level may be higher for farmers with lower productivity relative to the FCIC yield. Second, the heterogeneity of risk is reflected in the farmer's subjective variance or assessments of the probability of insurance effectiveness (ψ^β_i). Consideration of these differences then allows assessment of the extent to which insurance participation is influenced by adverse selection.

This model permits the investigation of moral hazard when the possibility of altering yields in the event of insurance participation is allowed. At the most simple level, moral hazard can be investigated by comparing the actual yield to the farmer's subjective expected yield (assuming that farmers' answers to the expected yield question reflect their standard production practices). More particularly, moral hazard can be investigated by considering relationships such as

$$y^*_i = y_i + f(\mu_i - \beta\,\overline{\mu}_i),$$

where y^*_i represents yield in the case of participation, y_i represents what would happen without insurance, and f describes the effect of moral hazard as a function of the level of insurance. A simple regression equation that captures the essence of this relationship (since both y_i and y^*_i cannot be observed simultaneously) is

$$\hat{y}_i = \mu_i + a \cdot (\mu_i - \beta \, \bar{\mu}_i)\theta,$$

where $\hat{y}_i$ is the observed yield regardless of insurance participation, a is a parameter to be estimated, and q is an indicator variable for insurance participation ($\theta = 1$ if insured and $\theta = 0$ if not).

8.6 The Survey Instruments

In this section, the survey instruments that were used to develop the data set are discussed briefly. Little detail is given regarding the FCRS since it is described elsewhere (Economic Research Service 1989). Somewhat more detail is given regarding the CATI survey.

The FCRS is an annual survey that is undertaken by the National Agricultural Statistical Service for the Economic Research Service to assess costs of production and returns in agriculture. It is a complex survey of farms nationwide with around 12,000 farmers included in each year. The same farmers are not included each year. This study relies on a subset of cash grain farmers in the 1988 FCRS survey who grew corn, grain sorghum, wheat, soybeans, cotton or rice. The questions cover acreage, production, sales, and individual variable input costs for 17 individual crops and two other broad crop groups. Similar information is obtained for livestock activities. Purchase, construction, and depreciation of capital equipment and buildings is also included. Finally, the data include financial and demographic data on the farmer such as assets and liabilities, off-farm income, age, and education. Of the 3208 such observations in the 1988 FCRS, only 2665 were usable due to incomplete responses.

The CATI survey is a telephone survey which is designed to follow the FCRS to obtain additional information for the analysis of insurance participation. The general questions in this survey focus on the farmer's insurance decision, and experience with disaster assistance. The bulk of this survey focuses on crop by crop questions assessing the stochastic distribution of yields on the farm and the probability of an insurance indemnity under each type of insurance contract. For each crop, questions are asked regarding expected average yield on the farm, the chances of an average farm yield of at least 50, 65, and 75 percent of the expected average yield, the ASCS Proven Yield, and the APH Yield.[4] The crop-specific questions are repeated for the irrigated and non-irrigated cases. Of the 2665 usable FCRS farms, relatively complete CATI observations were obtained on 1934 farms. This corresponds to a response rate of 72.6 percent.

Some inconsistencies in the data from the two sources further reduced the number of observations available for statistical analysis. Of the 1934 usable FCRS/CATI observations, 50 (2.6 percent) were excluded because the FCRS indicated participation in federal crop insurance and the CATI did not, and 85 (4.4 percent) were excluded because the CATI indicated participation in federal crop insurance and the FCRS did not.

The data files of the FCIC on actual federal crop insurance policies purchased by farmers were used to obtain data on insurance premiums and insured levels. In merging this data with the FCRS/CATI data, an additional 12 observations (0.6 percent) were excluded because the FCIC data indicated a positive premium and the FCRS indicated no federal crop insurance. No such

---

[4]The farmer's subjective variance of yield is not a term that can be requested directly from the farmer. First, it can be estimated by assuming a particular parametric form (normality) and ascertaining the farmer's likely worst annual average yield over a given period of years. For example, the likely worst yield in six years is approximately one standard deviation below the mean under normality. Second, the farmer's responses on the probability of 50, 65, and 75 percent of expected yield (along with the associated yield levels) can be used as 3 observations in a probit regression estimating the variance. Both methods were used here although the results are reported only for the latter approach. The two approaches produced similar results.

inconsistencies were found with the CATI data. The most serious problem was for those observations where the FCRS and CATI data indicated the purchase of crop insurance but the National Agricultural Statistical Service was unable to identify a corresponding FCIC insurance record. This occurred for 186 observations (9.6 percent). More seriously, these exclusions all came out of the insured component of the sample which is already thin because of low participation in the crop insurance program. Finally, an additional 9 observations (0.5 percent) were excluded because the FCIC data indicated a premium for a specific crop for which the FCRS data indicated no acreage. All of these problems reduced the sample of 1934 observations to 1592 usable observations.

Of the 1592 observations, 967 included corn acreage for which 123 farms were insured, 221 included grain sorghum acreage for which 25 were insured, 838 included wheat acreage for which 97 were insured, 738 included soybean acreage for which 97 were insured, 168 included cotton acreage for which 32 were insured, and 23 included rice acreage for which 3 were insured. For the purposes of this paper, the cotton and rice samples were not analyzed.

An additional consideration in evaluating the empirical results presented below is the complex survey design. Because the FCRS sample is not a random sample, consideration of weighting factors is necessary for extrapolating the results to aggregate conclusions for the farm sector. Also, the standard errors and significance levels for the regressions are valid under a narrower set of assumptions than usual.

8.7 The Empirical Results

The empirical results reported here can be grouped into three broad categories. First, the results of an investigation into the extent to which federal crop insurance participation reflects adverse selection are presented. Second, results are presented showing the extent to which the effects of moral hazard are observed. Finally, regression results explaining observed insurance participation decisions are reported.

8.7.1 Analysis of Adverse Selection. *Heterogeneity of Average Yields by Crop.* The presence of adverse selection in the data is investigated along the lines indicated near the end of Section 8.5 above. The first column of Table 2 gives the difference between the farmer's expected yield and the FCIC yield used for insurance purposes by crop and by insurance decision. This comparison is in the spirit of the discussion surrounding Fig. 3. These results show that insured sorghum farmers have a lower expected yield relative to the insurance yield than do uninsured sorghum farmers. This reflects an adverse selection problem. However, the difference is only about 1.3 bushels per acre. For wheat, the difference is also about 1.3 bushels and for soybeans the difference is a little over two bushels. In the case of corn, insured farmers actually have a higher expected yield relative to the insurance yield than do uninsured farmers. This is inconsistent with adverse selection due to heterogeneity of average yields. Furthermore, the difference is almost four bushels per acre.

The conclusion is that heterogeneity of average yields by crop is a source of adverse selection in federal crop insurance participation for some crops although perhaps not an important source. This conclusion may not be surprising given that most of the FCIC efforts to tailor the insurance program to local conditions are based on average yields (the first moment of stochastic yield distributions). For example, if all farms had sufficient approved production histories from which to calculate APH yields, then the FCIC yields used for insurance purposes should order farms correctly according to their heterogeneity of average yields. The observed differences here for sorghum, wheat, and soybeans are probably due to the farmers with lower average yields not keeping the records to generate APH yields.

Table 2. Evidence of adverse selection: differences in farmers' subjective distributions of yields by insurance decision[a]

Crop	Insurance Participation Decision	Difference in Mean Yields[b] (bushels)	Farmer Subjective Variance (bushels)	Probability of Indemnity 50% level	65% level	75% level
Corn	Insured	22.02	3709.50	0.13	0.18	0.23
		(24.38)	(4077.38)	(0.15)	(0.15)	(0.17)
		n = 108	n = 68	n = 98	n = 88	n = 82
	Not Insured	18.39	2575.88	0.06	0.12	0.19
		(23.04)	(3339.80)	(0.10)	(0.14)	(0.15)
		n = 670	n = 492	n = 642	n = 621	n = 587
Sorghum	Insured	8.96	3169.11	0.12	0.21	0.30
		(18.34)	(3582.65)	(0.11)	(0.14)	(0.19)
		n = 16	n = 9	n = 12	n = 12	n = 12
	Not Insured	10.25	1505.85	0.12	0.19	0.26
		(17.45)	(2374.24)	(0.16)	(0.16)	(0.16)
		n = 120	n = 92	n = 113	n = 105	n = 91
Wheat	Insured	3.66	373.44	0.12	0.16	0.25
		(9.50)	(651.46)	(0.15)	(0.15)	(0.16)
		n = 72	n = 54	n = 78	n = 71	n = 67
	Not Insured	4.95	603.00	0.07	0.12	0.18
		(11.94)	(1757.41)	(0.12)	(0.14)	(0.15)
		n = 552	n = 359	n = 555	n = 529	n = 496
Soybeans	Insured	5.00	257.56	0.08	0.16	0.25
		(7.46)	(285.11)	(0.10)	(0.14)	(0.15)
		n = 92	n = 69	n = 83	n = 82	n = 72
	Not Insured	7.14	279.78	0.07	0.13	0.21
		(8.88)	(423.28)	(0.12)	(0.14)	(0.15)
		n = 415	n = 383	n = 498	n = 483	n = 458

[a]The numbers in parentheses are standard deviations and n is the number of observations.

[b]The difference in means is the farmer's subjective mean yield minus the insurance yield.

Source: Calculations with the FCRS, CATI, and FCIC data.

Heterogeneity of Average Yields Across Crops. An important point suggested by the first column of Table 2, however, is that the FCIC yields are uniformly lower than farmers' expected yields. For corn, the FCIC yields used for insurance purposes are 18-22 bushels below farmers' expectations. That FCIC yields are consistently below farmers' expectations can be explained by the fact that FCIC yields are based on rather long yield histories and yields are generally increasing over time. Thus, FCIC yields may lag behind actual yields. For this reason, some have suggested the need for a trend adjustment in the APH yields (Skees 1988).

National average feed grain yields over the last decade have increased by an average of about 5 percent per year so a ten year historical average can be about 25 percent lower than current yield expectations. This is consistent with both the corn and sorghum yield differences in Table 2. By contrast, national average wheat yields over the last decade have increased by about 1 percent per year so a ten year average would be roughly 5 percent below current yield expectations. Similarly, national average soybean yields have increased by about 1.7 percent per year making a 10 year historical average about 8 or 9 percent lower. Again, this is roughly consistent with the yield differences in Table 2.

As a result the 50, 65, and 75 percent insurance levels are actually much lower as a percentage of current expected yields. In effect, an insurance indemnity cannot be collected unless the yield loss is considerably greater than 50, 35, or 25 percent, respectively. Furthermore, because of the different rates of technological growth of yields among crops, the failure by federal crop insurance to take account of yield growth results in an inequity among producers of different crops. Using the growth rates cited above, a corn farmer insured at the 50 percent level would have to experience a yield loss of 60 percent of current expected yield to collect an indemnity whereas a wheat producer would only have to experience a 52.4 percent yield loss. To the extent that this phenomenon explains the higher participation in, say, wheat crop insurance compared to corn crop insurance, this points out a form of adverse selection that exists across crops. This is a form of heterogeneity of average yields which FCIC methods do not reflect well.

Heterogeneity of Risk. The heterogeneity of risk is examined further in the remaining columns of Table 2. The second column gives the average farmer's subjective variance of yield by crop and by insurance participation decision. The subjective variances are estimated for each farmer by means of a three point probit regression of the paired observations $(\beta\mu_i, \psi^{\beta}_i)$, $\beta = .5,.65,.75$, given a mean of μ_i. These results are qualitatively similar to the case where variances among farmers are compared on the basis of the square of actual minus expected yield.

The comparison in the second column of Table 2 is in the spirit of the discussion surrounding Fig. 2. The remaining three columns give averages of farmer's probabilities of yield falling below 50, 65, and 75 percent of the farmer's expected yield.

The results of these comparisons are enlightening. In the case of corn, insuring farmers have a variance of yield that is 44 percent larger than non-insuring farmers. For sorghum, the corresponding percentage is 110. Only in the cases of wheat and soybeans is the variance of yields smaller for insured farmers. This suggests an important problem of adverse selection related to heterogeneity of risk for corn and sorghum. Note that differences of yield variability among farmers are not well reflected in FCIC methods of rate making; no consideration of yield variability in the approved production history is used by FCIC. Rates are set by county and vary by yield level rather than by yield variability.

A further check on these results is offered in Table 3 where the variation of actual yields about both the farmer's expected yield and the FCIC yield for insurance purposes is computed by crop and insurance participation decision. These results show a higher variation of yields about farmers' expectations for insured farmers than non-insured farmers for every crop except soybeans.

Turning to the last three columns of Table 2, the comparison reveals even more consistency. In every case, the insuring farmers have higher average probabilities of the insured event than do non-insuring farmers. Thus, the FCIC insures

Table 3. Evidence of adverse selection: yield variation by insurance participation decision[a]

Crop	Central Measure of Variance	Uninsured	Insured	Insured Level 50%	65%	75%
		bushels^2				
Corn	Farmer Expectation	2729.87 (3426.01) n = 710	3216.19 (3827.21) n = 108	1687.47 (976.23) n = 3	3489.78 (4173.06) n = 74	2801.56 (3133.74) n = 28
	Insurance Yield	1690.23 (2337.33) n = 704	1868.66 (2396.93) n = 111	656.16 (521.35) n = 3	1987.57 (2608.14) n = 76	1759.22 (1997.19) n = 29
Grain Sorghum	Farmer Expectation	557.76 (1032.63) n = 127	1177.49 (1631.81) n = 16	2500.00 (0) n = 1	888.10 (1158.71) n = 12	1894.21 (3230.95) n = 3
	Insurance Yield	708.31 (1312.12) n = 139	915.77 (2246.24) n = 21	248.06 (0) n = 1	886.25 (2396.09) n = 17	1305.64 (2016.75) n = 3
Wheat	Farmer Expectation	201.26 (375.37) n = 563	290.99 (273.97) n = 72	180.50 (255.27) n = 2	306.11 (271.17) n = 49	266.24 (290.36) n = 21
	Insurance Yield	235.16 (413.62) n = 643	233.78 (279.18) n = 88	113.19 (158.41) n = 2	238.30 (254.07) n = 61	232.42 (344.75) n = 25
Soybeans	Farmer Expectation	199.23 (271.66) n = 543	156.16 (192.02) n = 94	100.00 (0) n = 1	156.66 (186.11) n = 57	158.67 (208.80) n = 35
	Insurance Yield	146.09 (352.87) n = 435	120.63 (145.55) n = 93	324.00 (0) n = 1	96.62 (95.53) n = 56	155.19 (197.55) n = 35

[a]The numbers in parentheses are standard deviations and n is the number of observations.

Source: Calculations with the FCRS, CATI, and FCIC data.

farmers with higher probabilities of indemnity payments. These results imply that federal crop insurance is catering to high risk farmers.

Heterogeneity of Other Factors. Other factors which are potential causes of heterogeneity of risk among farms include farm size, pooling of risk across larger crop acreage, and irrigation. To consider these effects together with the role of mean and variance of yield, regression analysis was used. The results for corn (which is the only such equation estimated thus far) are as follows:[5]

$$\underset{(4.5)}{SDYLD = 18.7} - \underset{(.0070)}{.0150}\ CORNAC + \underset{(.00302)}{.00867}\ CROPAC + \underset{(.038)}{.218}\ EXYLD$$

$$- \underset{(3.1)}{8.00}\ IRR + \underset{(3.20)}{8.63}\ INS \qquad R^2 = .089,\ \bar{R}^2 = .080,\ N = 534$$

where:

SDYLD = farmer's subjective standard deviation of yield for corn;

CORNAC = farmer's total corn acreage;

CROPAC = farmer's total crop acreage;

EXYLD = farmer's subjective expected yield;

IRR = a zero-one indicator variable of irrigation; and

INS = a zero-one indicator variable of federal crop insurance.

---

[5]Although these results suggest the possibility that significance of acreage is prevented by a high correlation between corn acreage and total acreage, this is not the case. For example, when total acreage is dropped from the equation the significance of corn acreage is lower than in the results presented here.

These results imply that farms that are more specialized in corn production (larger corn acreage) tend to have lower risk. On the other hand, farms that are less specialized (larger total acreage for a given corn acreage) tend to have more risky corn yields. A possible reason for this result is that farms with more corn acreage may be better able to pool risk across different corn tracts while farms that are larger overall may have less need to diversify corn growing conditions since the crop mix is more diversified.

The results imply that other variables also have important effects. Irrigation has a significant negative sign indicating that irrigated production is not only relatively but absolutely less risky. The significant positive sign on the insurance variable indicates an adverse selection problem that occurs even after taking all of the other variables into account. Those who insure have a standard deviation of yields almost 9 bushels per acre higher than those who do not, ceteris paribus.

The results show a highly significant positive relationship between the yield expectation and the standard deviation. Driscoll (1985) notes that the FCIC historically has developed rates using the assumption that relative risk of crop yields is constant across farms, that is, the standard deviation of yield is proportional to expected yield. Skees and Reed (1986) alternatively argue on the basis of farm business records in Kentucky and Illinois that the standard deviation does not vary with expected yield. The results here present strong evidence that the standard deviation of yield is not constant with respect to expected yield.

To examine the FCIC position, another regression similar to the one above was run using the coefficient of variation (farmer's subjective standard deviation divided by the farmer's expected yield) as the dependent variable. The results are as follows:

$$\underset{}{\text{CVYLD}} = \underset{(.044)}{.597} - \underset{(.000069)}{.000142}\ \text{CORNAC} + \underset{(.0000297)}{.0000798}\ \text{CROPAC} - \underset{(.00038)}{.00177}\ \text{EXYLD}$$

$$- \underset{(.0303)}{.0493}\ \text{IRR} + \underset{(.0314)}{.0694}\ \text{INS} \qquad R^2 = .078,\ \bar{R}^2 = .069,\ N = 534$$

where

CVYLD = farmer's subjective coefficient of variation for corn yield.

Here, the hypothesis of constant relative risk (no effect of expected yield on the coefficient of variation) is strongly refuted. Apparently, risk is not constant either relatively or absolutely but increases in expectations somewhere between these two extremes. This is consistent with better management or higher quality soils contributing to the ability to control yield variations; nevertheless higher expected yields are associated with higher variances of yield. Thus, while the results here do not support the conclusions of Skees and Reed (1986), they do not suggest that the FCIC method of basing rates on an assumption of constant relative risk is sufficient to avoid adverse selection.

While these regressions show that yield risk varies in a plausible way among farms, one must bear in mind that the FCIC attempts to correct for heterogeneity of risk by varying insurance premiums by county and within county by yield. This relationship has not been given sufficient consideration. However, Table 4 gives some interesting evidence that premiums do not account adequately for the differences. The dispersion of both expected yields and the standard deviation of yields relative to the insurance premium is great.

To investigate the extent to which FCIC premiums are tailored to explainable individual risk levels, the FCIC premium was added to the estimated equations above, but it explained little variation in addition to the other variables. When the FCIC premium is added to the SDYLD equation, its t-ratio is only -.85, and in the CVYLD equation is only -.82. This implies that the FCIC premium structure does not account for several of the factors that cause risk to vary significantly among farms. The presence of other significant variables implies that other factors should be considered when tailoring insurance rates to individual farms. On the other hand, the low R^2 statistics for these equations suggest that

Table 4. Extent of diversity among farms and its reflection in FCIC insurance premiums for corn

Variable	Observed Sample Density							
Farmer's Expected Yield	30-50	50-70	70-90	90-110	110-130	130-150	150-170	170+
Sample Percentage	0.9	2.7	10.6	22.7	23.3	20.0	15.3	4.5
Expectation Relative to Premium	6-24	24-42	42-60	60-78	78-96	96-114	114-132	132+
Sample Percentage	28.0	37.6	16.8	11.5	5.5	0.2	0.0	0.4
Farmer's Standard Deviation	8-25	25-42	42-59	59-76	76-93	93-110	110-127	127+
Sample Percentage	16.4	37.8	26.4	9.1	4.0	2.3	2.3	1.7
Standard Deviation/ Premium	0-10	10-20	20-30	30-40	40-50	50-60	60-70	70-80
Sample Percentage	44.8	31.6	14.5	4.9	2.3	1.1	0.2	0.6
Coefficient of Variation	.10-.25	.25-.40	.40-.55	.55-.70	.70-.85	.85-1.0	1.0-1.15	1.15+
Sample Percentage	24.4	42.3	16.1	7.9	34.4	2.3	3.2	0.4

Source: Calculations with the FCRS, CATI, and FCIC data.

the adverse selection problem may not be solved easily.

8.7.2 Analysis of Moral Hazard. *Comparison of Actual Yields with Expected Yields.* The presence of moral hazard is investigated along the lines indicated near the end of Section 8.5. First, the actual yield is compared to the farmer's expected yield. The assumption here is that even though farmers may apply fewer productive inputs or take somewhat less managerial care when insured, they will not consciously incorporate these considerations into their responses to questions about yield expectations on their farms. Thus, systematic discrepancies between the two in the event of insurance are evidence of moral hazard. To the extent that farmers consciously reduce productive input use or managerial care and these effects are incorporated into their responses to the expected yield questions, the effects of moral hazard would be greater than estimated here.

Table 5 shows the results of comparing actual average yield to the farmer's expected yield by crop and insurance decision. All of the entries in Table 5 are negative. This is of little significance and merely reflects the fact that all farmers together tended to have a bad year. The important comparison is the magnitude of the difference in expectations between non-insured and insured farmers.

The results imply that yields compared to expectations were lower among insured farmers than noninsured farmers for all crops except soybeans (compare the first two columns of Table 5). Insured corn farmers tended to make 2.5 bushels per acre less (compared to expectations) than non-insured farmers. Given the magnitude of corn yields, this is less than a 3 percent effect. For wheat, the difference is about 9 bushels per acre. Compared to the expected yield levels for wheat, this amounts to a moral hazard effect greater than 20 percent. The grain sorghum difference is about 12 bushels per acre which is also approximately a 20 percent effect. For soybeans where the sign of the difference is not consistent with moral hazard, the difference is only about 1.6 bushels per acre or about 5 percent. These results suggest that the FCIC faces a serious problem of moral hazard for some crops. One possible explanation for the larger effect for sorghum and wheat than for corn and soybeans is that soil conditions tend to be

Table 5. Evidence of moral hazard: relationship of actual yield to expected yield by insurance participation decision[a]

Crop	Sample Average for Uninsured	Sample Average for Insured	Insured Level 50%	Insured Level 65%	Insured Level 75%
			bushels		
Corn	-39.19	-41.75	-40.02	-42.12	-41.70
	(34.57)	(38.55)	(11.33)	(41.69)	(33.18)
	n = 710	n = 108	n = 3	n = 74	n = 28
Grain	-9.99	-22.31	-50.00	-19.39	-24.76
Sorghum	(21.48)	(26.92)	(0)	(23.63)	(43.83)
	n = 127	n = 16	n = 1	n = 12	n = 3
Wheat	-1.39	-10.51	-9.50	-10.84	-9.82
	(14.12)	(13.52)	(13.43)	(13.87)	(13.34)
	n = 563	n = 72	n = 2	n = 49	n = 21
Soybeans	-9.81	-8.18	-10.00	-7.82	-8.67
	(10.15)	(9.49)	(0)	(9.85)	(9.27)
	n = 543	n = 94	n = 1	n = 57	n = 35

[a]In each case, the table reports the average across the sample of the farmer's actual yield minus the farmer's subjective expected yield. The numbers in parentheses are standard deviations and n is the number of observations.

Source: Calculations with the FCRS, CATI, and FCIC data.

less uniform within counties in the plains states that specialize in wheat and sorghum compared to the corn belt states that specialize in corn and soybeans. In addition, summer fallow is a common practice in the plains states. Thus, farmers may insure selectively when growing crops on poorer soil and not insure otherwise.

Regressions of Yield on Expected Yield and Insurance Level. Another approach to investigating the significance of moral hazard is the regression approach suggested near the end of Section 8.5. The results are reported in Table 6 for corn, sorghum, wheat, and soybeans. Three regressions are given in each case depending on whether the insured level is represented by a dummy variable reflecting the insurance decision, the insured yield level, or both.

Table 6. Evidence of moral hazard: regressions of actual yields on expected yield and level of insurance[a]

Crop	Coefficient of Insurance Indicator	Coefficient of Insurance Level[b]	R^2	n
Corn	-24.6	-.304	.682	778
	(11.3)	(.182)	(.681)	
	-41.8		.666	778
	(4.9)		(.665)	
		-.661	.680	778
		(.077)	(.680)	
Grain Sorghum	-2.01	-.727	.885	136
	(11.03)	(.332)	(.853)	
	-22.3		.861	136
	(6.0)		(.860)	
		-.778	.885	136
		(.179)	(.854)	
Wheat	-3.62	-.477	.913	624
	(3.17)	(.187)	(.913)	
	-10.5		.912	635
	(1.7)		(.912)	
		-.659	.913	624
		(.097)	(.913)	
Soybeans	-2.67	-.363	.796	507
	(3.17)	(.195)	(.795)	
	-8.18		.781	507
	(1.40)		(.780)	
		.511	.796	507
		(.086)	(.796)	

[a]Numbers in parentheses are estimated standard deviations of regression coefficients except in the case of R^2 where they are adjusted R2's. The coefficient of expected yield is constrained to be 1 and no constant term is included.

[b]Note that the insurance level is defined by the difference in the farmer's expected yield and the insured yield (50, 65, or 75 percent of the FCIC yield) if insured and zero if not insured.

Source: Calculations with the FCRS, CATI, and FCIC data.

These results suggest that moral hazard is present and highly significant in all four crops (the lower significance of individual coefficients when both variables are included is due to multicollinearity). The estimated coefficients are negative in all cases and an implied t-ratio exceeding 3.7 (in absolute value) is obtained for every crop in the regressions, eliminating multicollinearity. These results suggest the need to review FCIC monitoring procedures. With significant moral hazard problems among participants, it may be that farmers who do not reduce productive inputs or managerial effort when insuring find insurance participation unprofitable.

Explanation of Participation Decisions. Consider next the explanation of observed insurance decisions. This is done by regressing the insurance participation decision on the mean and variance effects of participation in a probit framework following equation (1). In particular, the model is of the form

$$P(\theta = 1) = f(\Delta\pi_i + \phi_i\Delta\sigma_i),$$

where $\Delta\pi_i = \bar{\pi}^*_i - \bar{\pi}_i$ and $\Delta\sigma_i = (\sigma^*_\pi - \sigma_\pi)A_i$.

Note that these right-hand-side variables depend on the choice of insured price and yield (α and β). One approach to this problem is to estimate a multinomial probit model with ten possible choices (noninsurance or insurance at each of the 9 price and yield combinations). However, the various insurance options are highly correlated so greater clarity is likely to be obtained in comparing the simple options of insurance and noninsurance. Thus, one set of insurance parameters has to be selected for each observation. For farmers who purchased insurance, the insured price and yield levels are observed while for uninsured farmers they must be estimated. To make the observations more comparable, however, the choice of insurance parameters was estimated for both insurers and noninsurers. This was done in several ways. First, the insurance parameters generating the largest expected profit per acre were selected. Second, the insurance choice was analyzed assuming only the middle yield and high price options were available (the most common choices). Third, the best insurance parameters were chosen, based on the certainty equivalent where the risk aversion coefficient estimated by the first approach was used to evaluate the certainty equivalent. The results associated with all three approaches were similar so only the results of the first approach are reported here.

The risk aversion coefficient is represented by $\phi_i = B'Z$ where B is a vector of unknown coefficients and Z is a vector of demographic and financial variables explaining risk aversion (see Calvin (1988) for a similar analysis of risk and the discrete/continuous choice of commodity program participation/acreage allocation). The resulting probit regression is of the insurance participation decision on $\Delta\pi$, $\Delta\sigma$, $\Delta\sigma Z_i$, $\Delta\sigma Z_2$, ...

The Z_i variables used for this purpose are as follows:

REQU = a dummy variable reflecting whether the farmer is required to buy insurance because of earlier program choices (1 if required and 0 otherwise);

OFFY = off-farm income;

PAST = past experience with disaster assistance programs (1 if disaster assistance was received since 1979 and 0 otherwise);

WLTH = wealth as reflected by farm assets less farm debts;

and

EDUC = education level of farm operator (1 if less than high school graduation, 2 if completed high school, 3 if some college, 4 if completed college, and 5 if beyond college).

The requirement variable is included to reflect the fact that some farmers who purchased insurance may not have done so if not required (which implies a negative effect of the term REQU $\cdot \Delta\sigma$). The off-farm income variable is included to investigate the diversifying role of off-farm labor. Farmers with assured income from other sources may be less concerned with risk (which implies a positive effect of the term OFFY $\cdot \Delta\sigma$) while, on the other hand, farmers who seek off-farm income may be those who are more risk averse (which implies a negative effect).

The past experience variable is included to investigate the role of disaster assistance perceptions. It has been argued that farmers who have received disaster assistance in the past tend to discount the variance reduction otherwise associated with insurance (which implies a positive effect of the term PAST $\cdot \Delta\sigma$). This represents the potential undermining effect on the crop insurance program of Congress passing ad hoc disaster assistance bills. On the other hand, farmers who have had more experience with disaster may be more aware of risk and place a heavier premium on its avoidance (which implies a negative effect).

The wealth variable is included to represent the dependence of risk aversion on wealth. With decreasing absolute risk aversion, more wealthy farmers have lower absolute risk aversion which implies a positive effect of the term WLTH $\cdot \Delta\sigma$.

The education term will have a positive effect on risk aversion if educated farmers know better how to rely on various vehicles other than insurance for laying off risk and have better potential for generating income elsewhere in the event of a disaster. On the other hand, the education term will have a negative effect if more educated farmers discover how to better incorporate insurance into their overall risk management strategy.

The results for corn are presented in Table 7. The first regression assumes that absolute risk aversion is the same among all farmers. The results are poor. Although the signs of the mean and variance effects are consistent with theory, the significance of coefficients is low and the overall explanation and significance is very low.

Table 7. Explanation of insurance decisions for corn: logit regression results[a]

Variable	Regression 1	Regression 2	Regression 3
Constant	-1.19 (.09)	-1.32 (.10)	-1.31 (.10)
$\Delta\pi/10$	.109 (.071)	1.19 (.76)	.139 (.079)
$\Delta\sigma/10^6$	-.008 (.094)	.162 .135	.338 (.221)
REQU · $\Delta\sigma/10^5$		-1.43 (.80)	-1.41 (.79)
OFFY · $\Delta\sigma/10^{11}$		-.745 (.288	-.734 (.304)
PAST · $\Delta\sigma/10^6$		-.258 (.122)	-.260 (.129)
WLTH · $\Delta\sigma/10^{11}$		.261 (.191)	.273 (.196)
EDUC · $\Delta\sigma/10^6$			-5.39 (5.10)
Likelihood Ratio Test	4.11 (2 df)	45.7 (6 df)	46.9 (7 df)
McFadden R^2	.013	.147	.151

[a]Numbers in parentheses are estimated standard errors except in the case of the likelihood ratio test where they give degrees of freedom for the test.

Source: Calculations with the FCRS, CATI, and FCIC data.

When risk aversion is allowed to depend on off-farm income, past experience, and wealth, and the effect of required purchases is considered as in the second regression, the R^2 statistic improves by an order of magnitude. Furthermore, the significance of individual coefficients improves substantially and plausibility is maintained. The estimated mean effect of insurance on profitability has a positive sign, as it should. While the estimated coefficient of the raw change in variance term is positive, the overall effect of the change in variance, considering all terms including it, is negative at the mean of the data and in a majority of cases. A requirement to buy insurance has a positive estimated effect, as expected (since $\Delta\sigma < 0$). The estimated wealth effect is to reduce risk aversion and thus make insurance participation less likely.

Turning to the a priori ambiguous terms, estimates show that off-farm income increases the likelihood of insurance participation. This is inconsistent with the explanation whereby off-farm income is an alternative to insurance participation and thus has a negative effect on participation. Rather, the results indicate that off-farm income and insurance participation tend to occur together as would be the case when more risk averse individuals tend to pursue both alternatives simultaneously.[6]

In the case of past experience with disaster assistance, the results are contrary to arguments that have been advanced against ad hoc disaster assistance bills. The past experience variable has a positive effect on the likelihood of insurance participation. This result is surprising and contrary to the arguments that the likelihood of ad hoc disaster assistance reduces crop insurance participation. Because this is a critical result, it was investigated in some detail. Apparently, the sign of this term in the regression is quite stable with respect to equation specification. Also, the raw correlation of all of the disaster assistance variables in the sample with the insurance participation decision are positive. Several explanations are possible. First, farmers who avail themselves of disaster assistance may tend to be those farmers who are more acquainted with all other tools available for laying off risk. Thus, the ones receiving disaster assistance may tend to be the ones who buy crop insurance. Second, with more disaster experience, a farmer may

---

[6]If off-farm labor and insurance are simultaneous decisions, then inclusion of off-farm income in the insurance participation regressions may cause simultaneous equation bias in the estimates. Future research will consider this simultaneity more carefully.

tend to give more weight to the possibility of disaster and thus become more interested in crop insurance. In any case, the results do not support the argument that disaster assistance programs are an important explanation for low participation in federal crop insurance.

The third estimated equation in Table 7 considers additionally the effects of education. It shows that education tends to increase the likelihood of insurance participation. This is consistent with the argument that more educated farmers are better able to incorporate insurance into their overall risk management strategies.

While the overall fit obtained in the estimated equations in Table 7 is low, one must bear in mind the nature of the sample. An R^2 in the neighborhood of .15 is significant given the large number of cross section observations, the truly micro level of the observations (county data average out a lot of unexplainable variation), and the extensive diversity that exists among producers and insurance marketing programs across the entire nation.

8.8 Conclusions

This paper examines a number of issues surrounding participation in federal crop insurance programs. As noted, a number of important reservations must be attached to the estimates presented here. Nevertheless, some interesting preliminary conclusions emerge.

First, the data suggest a considerable degree of adverse selection. However, this adverse selection is largely related to risk and factors other than differences in average yields which is the main factor on which the FCIC program focuses. Although the percentage points applicable for yield insurance (50, 65, or 75) do not vary among farms, the FCIC attempts to correct for heterogeneity of risk by varying insurance premiums by county and within county by yield. However, these premiums do not account adequately for the differences that are explainable on the basis of readily observed information. Thus, adverse selection due to heterogeneity of risk and other factors among farms appears to be an important explanation of low participation rates. A significant share of these problems can be eliminated by more carefully tailoring insurance premiums to farm size, crop acreage, irrigation, and farm-specific production conditions. Inequities in the program among the producers of different crops are also apparently important. These problems can be eliminated by building trend adjustments into APH yields.

Second, the evidence on moral hazard is remarkable. Substantial losses are occurring, apparently due to the modification of behavior in the event of insurance purchase. This evidence suggests that heavy government subsidization will continue to be necessary unless monitoring can be improved greatly.

Finally, the participation regressions show that limited explanation of the insurance participation decisions is possible. Apparently, the explainable variation is well represented by standard concepts of risk averse behavior once the role of various forms of heterogeneity among farms in determining the stochastic distribution of yields and risk aversion are taken into account. These various forms of heterogeneity include average yields, yield risk, farm size, diversification, financial structure, and insurance parameters. It remains for future research to consider the extent to which these observations can be exploited in improving federal crop insurance participation and the overall equity and efficiency effects of the federal crop insurance program.

References

Ahsan SA, Ali A, Kurian N (1982) Toward a theory of agricultural insurance. Amer J Agric Econ 64:520-529

Calvin L (1988) Distributional consequences of agricultural commodity policy. Unpublished Ph.D. dissertation, Univ California, Berkeley

Calvin L (1990) Participation in federal crop insurance. Paper presented at Southern Agric Econ Assoc Annual Meetings, Little Rock, Arkansas, February 4-7, 1990

Cawley Jr WA (1989) Report on federal crop insurance to Governor William Donald Schaefer. Maryland Dept Agriculture

Chambers RG (1989) Insurability and moral hazard in agricultural insurance markets. Amer J Agric Econ 71:604-616

Driscoll J (1985) Changes in ratemaking for federal crop insurance. Risk analysis for agricultural production firms: concepts, information requirements and policy issues. Proceedings of Southern Regional Project S-180. Dept Agric Econ, Michigan State Univ

Economic Research Service (1989) Financial characteristics of U.S. farmers. Agricultural Information Bulletin No. 579, U.S. Dept of Agriculture

Gardner BL, Kramer R (1986) Experience with crop insurance in the United States. In Hazell P, Pomareda C, Valdez A (eds) Crop insurance for agricultural development: issues and experience. Johns Hopkins Univ Press, Baltimore, Maryland, 195-222

Halcrow HG (1949) Actuarial structures for crop insurance. J Farm Econ 21:418-443

Hojjati B, Bockstael NE (1988) Modeling the demand for crop insurance. In Mapp HP (ed) Multiple peril crop insurance: a collection of empirical studies. Southern Cooperative Series Bulletin No. 334, Oklahoma State Univ, Stillwater

Just RE (1990) Modelling the interactive effect of alternative sets of policies on agricultural prices. In Winters LA, Sapsford D (eds) Primary commodity prices: economic models and policy. Cambridge Univ Press, Cambridge, England, 105-129

King R, Oamek G (1983) Risk management by Colorado dryland wheat farmers and the elimination of the disaster assistance program. Amer J Agric Econ 65:247-255

Lambert R (1983) Long-term contracts and moral hazard. Bell J Econ 14:441-452

Lee I (1953) Temperature insurance—an alternative to frost insurance in citrus. J Farm Econ 35:15-28

Miranda MJ (1990) Area-yield crop insurance reconsidered. Dept Agric Econ, Ohio State Univ

Nelson C, Loehman E (1987) Further toward a theory of agricultural insurance. Amer J Agric Econ 69:523-531

Nieuwoudt WL, Johnson SR, Womack AW, Bullock JB (1985) The demand for crop insurance. Agricultural Economics Report No. 1985-16, Dept Agric Econ, Univ Missouri, Columbia

Patrick G (1988) Mallee wheat farmers' demand for crop and rainfall insurance. Aust J Agric Econ 32:37-49

Quiggin J (1986) A note on the viability of rainfall insurance. Aust J Agric Econ 30:63-69

Quiggin J (1990) The optimal design of crop insurance. Dept Agric and Resource Econ, Univ Maryland

Rausser GC, Just RE (1981) Using models in policy formation. In Modeling agriculture for policy analysis in the 1980s. Kansas City Federal Reserve Bank, Kansas City, Missouri, 139-174

Rogerson W (1985) Repeated moral hazard. Econometrica 53:69-76

Skees JR (1988) Findings from extension special projects: evaluating multiple peril crop insurance. Paper presented at the National Workshop on Risk Management Strategies Utilizing Multiple Peril Crop Insurance, Kansas City, Missouri

Skees J, Reed M (1986) Rate-making for farm-level crop insurance: implications for adverse selection. Amer J Agric Econ 68:653-659

U.S. General Accounting Office (1989) Disaster assistance: crop insurance can provide assistance more effectively than other programs. Report to the Chairman, Committee on

Agriculture, U.S. House of Representatives, GAO/RCED-89-211, Washington, DC

U.S. General Accounting Office (1988) Crop insurance: participation in and costs associated with the federal crop insurance program. Report to the Chairman, Committee on Agriculture, U.S. House of Representatives, GAO/RCED-88-171BR, Washington, DC

U.S. General Accounting Office (1988) Crop insurance: program has merit, but FCIC should study ways to increase participation. Report to the Chairman, Committee on Agriculture, U.S. House of Representatives, GAO/RCED-88-211BR, Washington, DC

Zulauf C, Hedges D (1988) Disaster assistance for U.S. farmers: an overview of the current debate and two proposals. Dept Agric Econ and Rural Sociology, Ohio State Univ

Chapter 9

Crop Insurance and Crop Production: An Empirical Study of Moral Hazard and Adverse Selection

J. QUIGGIN[1], G. KARAGIANNIS[2] and J. STANTON[3]

Multiple risk crop insurance has not been very successful in the United States or elsewhere. As has been shown in other chapters in this book, no multiple risk crop insurance scheme has consistently earned enough premium income to cover payouts, much less administrative costs. The standard explanations for the failure of multiple risk crop insurance relate to problems of adverse selection and moral hazard. A number of theoretical models of these problems have been proposed (for example, Ahsan, Ali, and Kurian 1982, Quiggin 1986, Nelson and Loehman 1987, and Chambers 1989). However, there has been comparatively little empirical study of the problem. One reason may be that it is difficult, in practice, to distinguish between adverse selection and moral hazard.

A critical feature of the multiple risk crop insurance problem which differentiates it from many other forms of insurance, including those which have been the subject of most theoretical attention, is the fact that it takes place in a production context. Insured losses depend on the interaction of a wide range of exogenous variables such as rainfall and temperature, and endogenous variables including decisions on the level and timing of input use. By contrast, most insurance contracts and standard modelling of the insurance problem deal with a simple loss event in which some particular injury is sustained, or some item is

---

[1]Research School of Social Sciences, Australian National University, Canberra, Australia.

[2]University of Saskatchewan, Saskatoon, Saskatchewan, Canada.

[3]Department of Agricultural and Resource Economics, University of Maryland, College Park, Maryland 20742, USA.

destroyed or damaged. The actions of the insured person can normally be characterized fairly simply, for example, in terms of the effort expended to prevent or reduce loss. This characterization also applies to commercially successful insurance contracts offered to farmers, such as hail insurance.

The object of this chapter is to present a theoretical and empirical analysis of adverse selection and moral hazard. First, it is argued that the distinction between moral hazard and adverse selection, while enlightening in some contexts, is not really applicable in this framework. Theoretical predictions of the joint impact of these effects are derived from a model of choice under uncertainty. These predictions are tested in an empirical model based on a production function framework. A cross-section study using data from the U.S. Department of Agriculture's Farm Costs and Returns Survey is used.

9.1 Moral Hazard and Adverse Selection

It is usual in the literature to differentiate between adverse selection and moral hazard as follows. *Moral hazard* refers to the fact that the insured person's optimal decision may change as a result of taking out insurance. Because the insurance contract reduces the loss associated with the insured event, such changes in behavior will normally increase the probability of the insured event occurring. *Adverse selection* refers to the fact that people who are more likely to suffer the insured event will be more willing to insure at a given rate. If the insurance company cannot detect such people, then losses will occur.

While this distinction is useful for the purposes of analysis and exposition, it cannot be pushed too far. Suppose, for example that a farmer chooses to grow wheat on low-quality land and corn on high quality land, and to insure the wheat but not the corn. If the allocation of land to wheat is taken as given, this may be regarded as a case of adverse selection. The yield for wheat is likely to be lower than for corn, so it is rational to insure wheat but not corn. On the other hand, if the decision to insure the wheat crop is taken as given, this may be regarded as a case of moral hazard. The allocation of land by the insured farmer is such as to increase the probability of a payout. Since the allocation and insurance decisions must be regarded as co-determined, the distinction between adverse selection and moral hazard is inapplicable.

A more complex example arises if insurance may be taken out within some specified period after the date of planting. Suppose a farmer delays the planting date in order to gain more information on soil moisture, then makes a decision to plant without insurance if the outlook is favorable, plant with insurance if it is less favorable, and abstain from planting if it is highly unfavorable. Once again, the situation involves elements of both moral hazard and adverse selection. The farmer chooses to insure when the odds are favorable and not otherwise (adverse selection) and actions including the date of planting and the decision on whether to plant are altered because of the availability of insurance (moral hazard).[4]

A clear distinction between adverse selection and moral hazard can only be drawn in radically simplified models of farmers' insurance and production decisions. Such simple models are useful, but they may not prove a reliable guide to empirical analysis. The basic implication for production is the same for moral hazard and adverse selection. *Farmers who are insured will*

[4]The pejorative connotations of the term 'moral hazard' are unfortunate. In particular, these connotations appear to have led to some belief that analysts who regard moral hazard as a problem for multiple risk crop insurance are guilty of imputing immorality to American grain growers. The fact that an insured farmer chooses to plant a crop which would not be planted in the absence of insurance fits the theoretical category of moral hazard but would not be regarded as blamable conduct under most ethical codes. Indeed, if food production is regarded as a desirable goal regardless of its efficiency consequences (a view which appears to be held by most farmers and many non-farmers), then the farmer's actions would be regarded as commendable.

produce low yields more frequently than uninsured farmers with similar observed characteristics. This implication is tested here.

The critique of the adverse selection/moral hazard distinction drawn from the examples presented above may be presented in a more general way. Both 'moral hazard' and 'adverse selection' are implications of two assumptions which are central to modern economic theory—that individuals make rational choices in the light of the constraints facing them, and that information is costly. Only when the constraint and information sets take on particularly simple forms can specific choices be regarded as examples of either 'moral hazard' or 'adverse selection'.

9.2 The Model

The farmer's production decision may be modelled as follows. The farm has fixed inputs Z, and must choose observable variable inputs X and unobservable inputs (for example, operator's effort) θ. Farm output[5] Y depends on the input vector (Z, X, θ) and on random variables ε, representing exogenous fluctuations such as climate, and φ representing unobservable differences in land quality, farmer skills (referred to briefly as 'farm type'). The variable θ corresponds roughly to moral hazard in the usual models and the variable φ to adverse selection. The farmer must also decide whether or not to take out insurance.

The production function is assumed to be of the form

$$Y = f(Z, X, \theta)\, \eta(\phi\varepsilon) = F(Z, X, \theta, \phi) \qquad (1)$$

[5] As a simplification, it is assumed that the farm produces a single output. This assumption appears necessary to make the problem tractable but it does abstract from important aspects of the multiple risk crop insurance problem, most notably the option of insuring some crops but not others.

where f has the usual properties (positive first derivatives and cross derivatives, negative second derivatives) and $\eta(\phi\varepsilon)$ is a multiplicative shifter. The interaction between the random farm type and climatic variables, given by $\eta(\phi\varepsilon)$ is not specified in detail, but it is assumed that the lower the value of φ (that is, the worse the farm type) the lower is the mean and the greater is the variability of $\eta(\phi\varepsilon)$.

This production function differs from that proposed by Just and Pope (1979), who suggest a functional form which admits both risk-increasing and risk-reducing inputs. The main problem with the Pope-Just approach is that the derivation of input demands is complex when more than one input affects risk.

The profit function in the absence of insurance may be written

$$\pi = pY - w \cdot X - C(\theta) \tag{2}$$

where:

p is the of output price;

w is the vector of prices for observable inputs; and

$C(\theta)$ is a cost function for unobservable inputs.

The farmer seeks to maximize $E[U(\pi)]$ where U is a von Neumann-Morgenstern expected utility function.

The optimization problem is to choose X and θ so as maximize $E[U(\pi)]$. The first order conditions are

$$E[U'(\pi)(\partial R/\partial X - w)] = 0, \tag{3a}$$

$$E[U'(\pi)(\partial R/\partial \theta - C'(\theta))] = 0, \tag{3b}$$

where

$R = p \cdot Y$ is gross revenue,

$\partial R/\partial X = p\partial f/\partial X \ - \eta(\varphi\varepsilon)$,

$\partial R/\partial \theta = p\partial f/\partial \theta \ - \eta(\varphi\varepsilon)$.

The effect of insurance is to replace the revenue term $R = p \cdot Y$ in (2) with a modified revenue R^*, given by

$$R^* = p \cdot Y - \rho \quad \text{if} \quad Y \geq Y^* \tag{4}$$

$$= p^* \cdot Y^* - \rho. \quad \text{if} \quad Y < Y^*$$

where Y^* is the guaranteed yield, p^* is the price election and ρ is the premium. The marginal revenue product becomes

$$\partial R^*/\partial X = p\partial f/\partial X \ \eta(\varphi\varepsilon) \quad \text{if} \quad Y \geq Y^* \tag{5}$$

$$= 0 \quad \text{if} \quad Y < Y^*$$

It is straightforward to show that

(i) The lower is φ (the worse the farm type) the more profitable is any given insurance contract, for the farmer,

(ii) Given decreasing absolute risk aversion, there exists a φ^* such that insurance will be taken out if $\varphi \leq \varphi^*$, and not if $\varphi > \varphi^*$. This is the adverse selection effect.

(iii) Given decreasing absolute risk aversion, insurance leads to a reduction in the optimal levels of X and θ.

9.3 Model Estimation

The crucial problem for modelling is that θ and φ (as well as ε) are unobservable. However from the results (i)-(iii) above, the observed insurance decision, denoted δ, may be used as a proxy for

φ and θ. Hence it is possible to estimate a system consisting of a production function and input share equations

$$Y = f(X, Z, \delta) \tag{6}$$

$$X = h(Z, p, w, \delta) \tag{7}$$

The analysis above suggests the hypotheses that the coefficient on δ will be negative in both the production function and the input demand functions.

In the empirical analysis reported here, a simple Cobb-Douglas functional form is used for (1). While the Cobb-Douglas functional form imposes restrictive assumptions (such as unit elasticities of substitution), these relate to issues which are not of central concern in the present study. The advantages of the Cobb-Douglas form include the simplicity of the derived input demand equations, the robustness derived from the fact that it is a first-order approximation to an arbitrary functional form, the fact that it displays reasonable behavior for all input values, and its parsimony in parameters. These characteristics make it an appropriate choice for this study. Thus, the functions (6) and (7) may be estimated in the standard log-log and input share forms with the addition of a dummy indicating insurance

$$\ln Y = \alpha_0 + \sum_{i=1}^{n} \alpha_i \ln X_i + \sum_{i=1}^{n} \beta_j \ln Z_j + \delta_0, \tag{8a}$$

$$w_i X_i / p\, Y = \gamma_i + \delta_i. \tag{8b}$$

As noted above, the insurance dummy variables δ are expected to have negative signs. Under the assumption of profit maximization, the input demands and production function satisfy the cross-equation constraint $\gamma_i = \alpha_i$. Under the hypothesis of risk-aversion, assuming maximization of expected utility or some appropriate generalized functional, the equality constraint is replaced by an inequality $\gamma_i \leq \alpha_i$. Usage of the variable inputs is expected to be lower than the level which would equate marginal cost with expected marginal value product.

There may also be some interest in more traditional hypotheses associated with the Cobb-Douglas functional form, such as those concerning returns to scale. Since insurance is offered on a per acre basis, non-constant returns to scale will automatically generate adverse selection problems. If for example, there are increasing returns to scale over some range, small farms will have lower expected yields and will be more likely to take out insurance. Just and Calvin (this book) point out that if returns on different parts of a large farm are not perfectly correlated, the variance for the farm as a whole is likely to be lower, resulting in further adverse selection effects. If production is extended into the range of decreasing returns to scale (as would be expected on the basis of textbook micro theory) then the mean and variance effects will work in opposite directions. In the long-run, the availability of insurance will lead to sub-optimal scale decisions, so that moral hazard and adverse selection effects are intertwined.

9.4 Data

The data for this study were derived from the 1988 *Farm Costs and Returns Survey* undertaken by the National Agricultural Statistical Services. Over 4000 farmers were interviewed for the original survey. For the present study, attention was confined to a population of grain farmers sufficiently homogeneous to permit the estimation of a production function. A subset of 18 major grain producing states was chosen. Farmers who derived more than 15 percent of their income from either cotton or vegetable production, or allocated more than 10 percent of their land area to pasture were excluded. Along with editing and consistency checks, these restrictions resulted in a sample of 535 producers.

Estimation of the model specified above requires data on output, variable and fixed inputs and insurance status. Output is measured in value terms, in order to permit the aggregation of a heterogenous output mix. Output value was measured as the sum of gross receipts from farm crop sales, total production bonuses, Commodity Credit Corporation (CCC) loans and cash receipts from the sale of livestock products. For an accurate measure of livestock input, it would be desirable to include a measure of the change in livestock holdings, but this was not feasible because the only available measure (change in value of livestock) conflates quantity and value changes. This difficulty, as well as the desire to focus on cropping enterprises, was a motive for the exclusion of farmers with large livestock operations.

Two fixed inputs, land and capital, were included. Each of these inputs was measured in stock value terms. Capital is formed as the aggregate value of farm buildings (barns, silos, cribs, equipment shops, grain bins, storage sheds) and plant and machinery (trucks, tractors, machinery, tools and implements). Variable inputs were partitioned into five categories—labor, fertilizer, pesticides, energy and other. All were measured in value terms. The only input where this presented difficulties was labor. Information on operator and family labor inputs was available in hours worked. Hired labor was available as an expense item. In order to provide an aggregate wage measure, an imputed wage was adopted. The imputed wage for operators was $10/hour and for other labor $4/hour.

It is common to aggregate fertilizer and pesticides into a more general input category. Because of the primary concern with the risk characteristics of inputs, it was not clear whether this approach was appropriate for the present study. Pesticides are generally viewed as a risk-reducing input, and fertilizer as a risk-increasing input. However, testing revealed no significant loss in power from aggregating the two inputs. This aggregation had the side benefit of permitting the inclusion of certain observations with zero values in one of the two categories, usually pesticides. The need to have positive values for all inputs is a weakness of the Cobb-Douglas production function (one shared by many other popular functional forms). In the present study, this problem arose only in relation to the fertilizer and pesticide category.

Fertilizers are defined here to include expenses for seed and plants, lime, and soil conditioners. Pesticides include insecticides, herbicides and fungicides. The energy input includes fuel, motor oils, electricity, water and telephone. All other cash outlays are aggregated into the 'other' input.

Insurance status was defined by a dummy variable taking the value 1 if the farmer participated in the Federal Crop Insurance Corporation program in 1988 and zero otherwise. An alternative would be to employ a continuous variable reflecting total premiums paid or total coverage. This approach would have the advantage of capturing information on levels of insurance coverage and on whether farmers insured all or only some crops. However, because of variations in premium rates between counties and between farmers within counties, the extraction of this information would be difficult, and the introduction of errors would be likely.

Finally, state-level dummies were incorporated in the analysis. In the production function, this permits the incorporation of multiplicative differences in total factor productivity between states. Given the sampling rate, it would not be feasible to include more finely specified geographical dummies. However, a theoretically preferable approach would be to include information on soil characteristics and on climatic characteristics (both 'normal' characteristics and 1988 experience) in place of a simple geographical dummy.

9.5 Results

The system of equations given by (8) was estimated by Ordinary Least Squares. Re-estimation using Zellner's SUR technique did not result in significant changes in estimated coefficients. The main results are presented in Tables 1a and 1b. State-level dummy variables were included in the estimation, but are not reported. The results for the production function are discussed first (Table 1a).

Table 1a. Estimated Coefficient of a Cobb-Douglas Production Function: Cross Section, Data, U.S.A., 1988

Parameters		Estimated Coefficient	Standard Error
Intercept	a_0	-0.44	(0.40)
Land	β_1	0.11	(0.03)
Capital	β_2	0.11	(0.05)
FrtPest	a_1	0.36	(0.04)
Labor	a_2	0.08	(0.04)
Energy	a_3	0.16	(0.06)
Other	a_3	0.40	(0.04)
Insurance	δ_0	-0.10	(0.07)
	N = 535		
	R2 = 0.78		

The Cobb-Douglas production function is well-behaved, with all input variables being right-signed and significant at the 5 percent level. The R^2 was 0.78, which is very satisfactory for a cross-section equation. The coefficients add to 1.18, indicating weak economies of scale. The hypothesis of constant returns to scale cannot be rejected at standard levels of significance. The absence of scale economies is a necessary, though not a sufficient, condition for the successful operation of a crop insurance scheme based on expected yields per acre, determined on a geographical basis. If large farms have consistently higher yields per acre than

Table 1b. Estimated Factor Shares for a Cobb-Douglas Production Function: Cross Section Data, U.S.A., 1988

Parameters		Estimated Coefficient	Standard Error	Parameters	Estimated Coefficient	Standard Error
FrtPes	γ_1	0.32	(0.16)	δ_1	-0.03	(0.09)
Labor	γ_2	0.08	(0.04)	δ_2	-0.01	(0.02)
Energy	γ_3	0.16	(0.06)	δ_3	-0.03	(0.03)
Other	γ_4	0.40	(0.04)	δ_4	-0.05	(0.04)

small ones, this pricing approach generates adverse selection, with small farms choosing insurance and large ones choosing to self-insure. One problem which may be observed from the estimated coefficients is that the sum of the coefficients on the variable inputs is only slightly below one. The validity of the law of diminishing marginal returns depends on the requirement that the variable input coefficients should sum to less than one. This issue is not of central concern for the present study.

The coefficient on insurance is negative, but not statistically significant. This, however, is a case where statistical significance and economic significance are not equivalent. The estimated coefficient indicates that insured farmers in a given state produce 10 percent less gross output than non-insured farmers in the same state using the same inputs.

The impact of insurance in the factor share equations (Table 1b) is once again negative, but insignificant. The estimated coefficients indicate that insured farmers tend to use less variable inputs to produce a given output than do uninsured farmers. In combination with the previous observation that output for a given vector of inputs is lower for insured farmers, this implies that insured farmers use less variable inputs in relation to fixed inputs, and in particular less inputs per acre of land than do uninsured farmers. This result is conducive to a moral hazard explanation. However, as noted above, it is also consistent with adverse selection. The estimated reduction in cost share is generally of the order of 10 percent.

The problem of statistical significance means that individual coefficient estimates must be regarded as preliminary. However, the joint null hypothesis that all of the insurance coefficients are zero may be rejected, at the 5 percent level, in favor of the alternative hypothesis, based on the assumption of economic rationality (or equivalently, adverse selection and moral hazard) that all coefficients should be negative.

9.6 Insurance Implications

The results derived in the previous section are consistent with the predictions of economic models of moral hazard and adverse selection. They imply that insured farmers will have lower observed levels of variable inputs and lower total factor productivity than uninsured farmers. These two effects may be combined and their impact analyzed in a simulation model.

Normalize the system of equations (8) by setting $w = p_i = 1, \forall i$. (This was done in the econometric estimation by expressing output and all variable inputs in value terms.) Substituting from the share equations into the production function and cancelling common terms yields ρ, the ratio of expected output per acre for insured farmers to expected output per acre for uninsured farmers:

$$\ln\rho = \delta_0 + \sum_{i=1}^{n} \alpha_i \ln((\gamma_i + \delta_i)/\gamma_i), \qquad (9)$$

and substituting the estimated coefficients from Table 1a and 1b yields an estimated value $\ln \rho = -0.21$ or $\rho = 0.81$.

In order to determine the impact of such a differential in an insurance scheme, it is necessary to incorporate uncertainty, arising from variation across farms and over time. Because the insurance scheme operates on the basis of output per acre, it is natural to treat uncertainty in terms of the coefficient of variation. For a given distribution, such as the normal or lognormal, and an insurance policy which specifies a given insured yield level, the coefficient of variation determines both the probability of a payout and the expected value of payouts. One problem with the use of the coefficient of variation in the present study arises from the observation that insured farmers have lower expected yields per acre than uninsured farmers. If the coefficient of variation is the same for the two groups, the insured farmers will have a lower variance and standard deviation. Since, as shown above, both moral hazard and adverse selection imply that insured farmers should be characterized by high variability, the assumption of equal coefficients of variation seems inappropriate. It will be assumed here that insured and uninsured farmers have the same variance, implying a higher coefficient of variation for insured farmers.

A simple illustration of the insurance implications of adverse selection and moral hazard arises when the parameters of the insurance policy are set in an actuarially fair fashion on the basis of observations of uninsured farms. Table 2a gives the probability of payout and the actuarially fair premium for the uninsured groups, for various levels of coverage and coefficients of variation (CV) of 0.2 and 0.3. Table 2b gives the same information for a lognormal distribution. Tables 3a and 3b give the corresponding payout probabilities, expected payouts and loss ratios for a group with similar variance and 20 percent lower expected yields.

As an illustration, with a coefficient of variation of 0.3 and a guaranteed yield equal to 65 percent of the expected yield, an actuarially fair scheme would charge a premium equal to 1.7 percent of expected yield, and would be expected to make payments on about 12 percent of its policies, on average. If the same policy were offered to a group with a 20 percent lower expected output, payments would be made on 24 percent of policies and expected payments would be equal to 4.4 percent of the original expected yield. The loss ratio (the ratio of payouts to premiums) would exceed 2.5. This is somewhat worse than the observed loss ratio for the Federal Crop Insurance System, which was about 1.8 over the 1980's (Gardner, this book). If the premium were calculated in line with normal insurance practice to include a 10 to 20 percent margin for administrative costs, but still using the uninsured group as a basis, the observed loss ratio would fall to between 2.00 and 2.25.

As shown in the tables, the loss ratio arising from adverse selection and moral hazard is greater for the lognormal distribution than for the normal, and greater for lower levels of coverage. This is because, in these cases, the probability of a payout derived from observations of the uninsured population is very low, leading to low premiums.

Table 2a. Parameters of fair insurance schemes: normally distributed yields

	CV = 0.2		CV = 0.3	
Coverage level[a]	Expected payout[b]	Probability[c]	Expected payout	Probability
65	0.3	3.6	1.7	12.0
70	0.6	6.5	2.4	15.7
75	1.0	10.4	3.3	20.0

[a]As percentage of expected yield
[b]As percentage of expected yield
[c]Percent

Thus, yield differences between insured and uninsured groups of the magnitude estimated here pose a problem for the operation of a successful system of crop insurance. Rates high enough to yield a positive expected return for the currently insured group would be so high as to make insurance unattractive to all but extremely risk-averse members of the uninsured group.

In practice farmers do not fall into two discrete groups. Rather among farmers (and farms) with similar observed characteristics there will be a continuum of ability levels, soil quality and ability to adjust input mix in response to economic incentives. It may be that there is no insurance policy which will be attractive to a significant group of farmers while having premiums sufficiently high to yield a positive return on the inevitable bad risks.

Table 2b. Parameters of fair insurance schemes: log-normally distributed yields

	CV = 0.2		CV = 0.3	
Coverage level	Expected payout	Probability	Expected payout	Probability
65	0.1	1.8	0.8	9.4
70	0.2	4.5	1.4	14.5
75	0.6	9.3	2.2	21.0

Table 3a. Parameters of insurance schemes with mean yields reduced 20 percent: normally distributed yields

	CV = 0.2			CV = 0.3		
Coverage level[a]	Expected payout[b]	Probability[c]	Loss ratio[d]	Expected payout	Probability	Loss ratio
65	1.5	15.0	5.0	4.4	24.3	2.6
70	2.5	22.2	4.2	5.8	30.0	2.4
75	3.8	30.1	3.8	7.4	35.9	2.2

[a]As percentage of expected yield
[b]As percentage of expected yield
[c]Percent
[d]Ratio of payouts to premiums, determined from Table 2a or 2b

9.7 Policy Implications

The results presented in the previous sections indicate the difficulties associated with a system of crop insurance. They suggest that the losses observed in such schemes around the world cannot be attributed primarily to mismanagement or lack of private sector expertise, but are inherent in this type of insurance. Policies for the reform of crop insurance must be formulated in this light.

Two basic approaches to policy reform may be considered. The first remains within the framework of multiple risk crop

Table 3b. Parameters of insurance schemes with mean yields reduced 20 percent: log-normally distributed yields

	CV = 0.2			CV = 0.3		
Coverage level	Expected payout	Probability	Loss ratio	Expected payout	Probability	Loss ratio
65	1.2	15.6	12.0	3.4	26.5	4.3
70	2.1	23.6	8.5	5.0	34.7	3.6
75	3.5	32.9	5.8	7.0	42.3	3.2

insurance, and involves attempts to mitigate the problems of adverse selection and moral hazard. The second involves more or less fundamental departures from the principles of multiple risk crop insurance. Two examples of the first approach are attempts to overcome the problem of adverse selection using individual farm records and attempts to overcome moral hazard using deductibles and co-payments. Consideration of the examples presented above indicates that neither of these approaches is likely to be adequate.

Consider first the use of policies based on individual farm records. The longest history which could reasonably be obtained for any large group of farms is of the order of 10 years. Given a coefficient of variation of 30 percent, the relative standard error associated with a mean estimate based on 10 observations will be of the order of 10 percent. This is a lower bound estimate which would apply in the case when expected yields for all farms were stable over time, at least in relative terms. In fact, over any period of ten years, some farms will enjoy increases in efficiency and others will experience decline. It seems likely, then, that even with individual rating, the adverse selection problem is likely to remain of the same order of magnitude as that estimated in the present study.

Similar problems arise with attempts to resolve the moral hazard problem. Economic analysis of insurance provides conditions for the existence of an optimal contract involving deductibles. In the present case, however, even a 35 percent deductible appears to be too small to make insurance feasible. As a practical matter, it appears unlikely that any insurance scheme with a significantly larger deductible would be politically feasible, since the payouts would be insufficient to cover variable costs for many farmers. Informal evidence suggests that any insurance policy which does not 'make farmers whole' in this respect will be rejected as unfair, regardless of its actuarial properties. Equally importantly, the calculations presented here indicate that as the deductible becomes larger, the proportional difference between the payout probability for the currently insured and that for the uninsured group becomes larger. Thus, the device of increasing deductibles deals with the moral hazard problem only at the expense of increasing the severity of the adverse selection problem.

A further approach to the adverse selection problem would be an attempt to identify instrumental variables which could be used to estimate the risk status of different producers. Candidates would include educational status, farming experience and perhaps debt status. It seems likely however, that most of the differences between farms and farmers will not be captured by instrumental variables of this kind.

There are a number of alternative approaches to policy reform based on modifications of the standard multiple risk crop insurance model. Given the theoretical prediction, supported by the estimates presented here, that insured farmers will have lower yields than uninsured farmers with similar observed characteristics, a natural approach is to propose insurance schemes in which the payout is independent of individual yields. Examples of this approach include regional yield insurance schemes discussed by, among others, the Australian Industries Assistance Commission (1977) and Glauber et al. (this book), and rainfall insurance schemes discussed by Bardsley, Abey and Davenport (1984) and Quiggin (1986).

As is argued by Quiggin (this book), this approach will not, in general, be optimal. A preferred approach will typically involve a payment which is defined by the observed yield loss, in combination with some deductible, as in the standard multiple risk crop insurance schemes, but is contingent on the occurrence of some insured event. The insured event could be the failure of regional average yield to reach some predefined value, as in the Glauber et al. (this book) model, or a predefined climatic event, as in the rainfall insurance literature.

A final possibility is the abandonment of insurance altogether in favor of a system of disaster relief, either *ad hoc* or based on some formal criteria. A scheme based on formal criteria is essentially equivalent to free insurance. The choice between the two approaches depends on administrative simplicity and on the general problem of choice between subsidy and free provision of goods. An *ad hoc* scheme may approach the characteristics of a lump-sum transfer if eligibility is determined on essentially arbitrary criteria such as the political weight of representatives from an area experiencing low yields. However, precisely in this case, its risk-reducing and distributional benefits are likely to be small or negative.

9.8 Concluding Comments

The theory of insurance predicts systematic differences between insured and uninsured firms. These predictions may be tested by a variety of comparisons between the two groups. The present chapter is based on the view that the most relevant comparison for a system of production insurance is one based on an explicitly specified production function. A number of modifications of the approach adopted here may be worth considering.

The Cobb-Douglas function used in the estimation here is simple and robust. One modification to the analysis presented here would be based on the use of more sophisticated and flexible functional forms. In particular, it would be desirable to employ a more general system of input demand equations. Horowitz and Lichtenberg (this book) have investigated input demand, but not in a production function context. Because not all products are covered by crop insurance and because multi-output producers may choose to insure some but not all of their crops, it would also be desirable to extend the analysis presented here to include explicit modelling of multi-output production.

Finally, the data used here represent only a single year, 1988. All years are atypical, but the drought year of 1988 was more so than most. An extension from cross-section to panel data would be desirable for a number of reasons. First, it would improve the reliability of estimates. Second, the relationship between variation over time and variation between farmers is critical to the operation of insurance schemes. An insurance scheme will work well when the output of individual farmers is independently and identically distributed over time, and will work badly when farmers with similar observed characteristics have different output distributions or when output is highly correlated across farmers.

Despite these *caveats*, the estimates presented here support the view that farmers' production and insurance decisions are responsive to economic incentives, and that these incentives work in a way which undermines the viability of a multiple risk crop insurance scheme.

References

Ahsan S, Ali, Kurian (1982) Toward a theory of agricultural insurance. Amer J Agr Econ 64:520-529

Bardsley P (1986) A note on the viability of rainfall insurance—reply. Aust J Agric Econ 30:70-72

Bardsley P, Abey A, Davenport S (1984) The economics of insuring crops against drought. Aust J Agric Econ 28:1-14

Chambers R (1989) Insurability and moral hazard in agricultural insurance markets. Amer J Agric Econ 71:604-616

Gardner (this book)

Glauber et al. (this book)

Horowitz and Lichtenberg (this book)

AIAC (1977) Report on rural income fluctuations. AGPS, Canberra

Just and Calvin (this book)

Just R, Pope R (1979) Production function estimation and related risk considerations. Amer J Agric Econ 61:277-284

Nelson C, Loehman E (1987) Further toward a theory of agricultural insurance. Amer J Agric Econ 69:524-531

Quiggin J (1986) A note on the viability of rainfall insurance. Aust J Agric Econ 30:63-69

Quiggin (this book)

Chapter 10

Crop Insurance Decisions and Financial Characteristics of Farms

H. LEATHERS[1]

10.1 Introduction

To the consternation of economic theorists and agricultural policy makers, many farmers in the United States evince a hardy disinterest in crop insurance, even when that insurance appears to be "fair" or "more than fair" (that is, subsidized by the government). A variety of explanations have been advanced for this disinterest. One large category of explanations contains variations of the argument that farmers receive effective insurance through sources other than subsidized crop insurance. For example, a system of disaster payments during low yield years may obviate the need for crop insurance for many individual farmers. Likewise, farmers may avoid the consequences of low yields by avoiding the low yields through use of inputs such as pesticides and irrigation, thus making crop insurance unnecessary. (See Glauber, Harwood, and Miranda 1989, and Commission for the Improvement of the Federal Crop Insurance Program 1989 for an overview of crop insurance issues.)

This paper was developed in response to a suggestion that the terms of farm operating loans may provide a form of insurance to farmers, thus making insurance unnecessary. If operating loans are secured by the farmer's crop, then the farmer may have limited liability: when yields are high, the farmer will repay the loan and keep the surplus; but when yields are low, the farmer will lose the crop, but no additional repayment will be required. In this situation, the farmer might see nothing to be gained from crop insurance, if the indemnity payments in periods of low yield were

[1]Department of Agricultural and Resource Economics, University of Maryland, College Park, MD 20742, USA.

simply to be turned over to the lender as a condition of the loan. On the other hand, it would seem that in such circumstances, the lender might require the borrower to take out crop insurance as a condition of the loan.

In previous work the interactions between lenders and crop insurance have been investigated by constructing a hypothetical case study (Pflueger and Barry 1986) and using Monte Carlo simulation methods (Leatham, McCarl, and Richardson 1987). This latter paper is especially relevant to the current work, concluding: "the lender ... always prefer[s] crop insurance. This was especially true when yield variability led to farm failure" (p. 120). More recently, Calvin's (1990) paper modelling crop insurance participation includes some financial characteristics (notably debt/asset ratio) as explanatory variables.

The purpose of this paper is to explore what ramifications the financial situation of a farm has for the decision to insure crops. The first section presents survey results which show that about a fifth of all farmers who insure their crops do so at the insistence of a lender. The second section presents a theoretical model of loan contracts which suggests a trifurcation of the farm population into those who do not insure, those who voluntarily insure, and those who are required to insure. In the third section, empirical evidence is presented which supports the view that financial characteristics are an important factor in determining whether or not a farmer will insure crops. In the final section, some implications for policy are discussed.

10.2 Survey Results

The data in this paper are for a subsample of crop farms drawn from the Farm Costs and Returns Survey. In addition to the information picked up as part of the Farm Cost and Returns Survey, these farmers were asked questions concerning their participation in the crop insurance programs, and factors which might influence that participation. The Farm Costs and Returns

Survey is based on a "complex sample design" which is intended to insure that accurate descriptive statistics for the entire population of farmers can be developed from the survey results. (See Fuller 1984 or Kaplan and Francis 1979 for a description of the complex sample design and appropriate statistical methods.) Of the 2600 or so farms in the crop insurance subsample, complete and usable responses were available for over 1900.

The general breakdown of farms is shown in Table 1. Slightly more than a quarter of farms in the sample bought crop insurance in 1988. Of those who bought insurance about one-fifth did so as a requirement of a loan. Slightly less than half of those who were required to insure had that requirement imposed by a commercial bank or savings and loan. Slightly over a third were required to insure by the Farmers' Home Administration (FmHA), and about a fifth were required to insure by a lender in the cooperative Farm Credit System (FCS, comprised of Production Credit Associations and Federal Land Banks). (These numbers do not add up because some farmers reported that insurance was required by two types of lenders.)

Table 1. Distribution of farms in the crop insurance subsample of the Farm Costs and Returns Survey, 1988

Group	Number	Percent of Total	Percent of Insured
Total	1921	100	
No Crop Insurance	1390	73	
Crop Insurance	514	27	100
Required to Insure by			
Commercial Lenders	48	2.5	9
FmHA	36	1.9	7
FCS	18	0.9	3.5
Individuals	5	0.2	1.0
Life Insurance Companies	1	0.5	0.2

Average financial characteristics of farmers are shown in Table 2. The numbers reported here are raw data, not weighted by weights reflecting the sample design. We see that, on average, the non-insurers farm fewer acres, and have lower cash farm income and lower farm debt. Voluntary insurers, on average, farm more acres, have higher cash farm income and higher debt. The farmers who were required to insure have, on average, high debt loads, low net worth, and low farm income per acre. As we see in the next section, this profile of farmers is consistent with a model in which the structure of the farmer's loan contract influences his/her decision to insure crops.

Table 2. Average financial characteristics of farms in the crop insurance subsample by farm type

	All Farms	No Crop Insurance	Voluntary Insurance	Required Insurance
Number	1921	1390	414	100
Acres Operated	1431	1228	2086	1532
Net Cash Farm Income	$55,358	$49,721	$80,451	$41,134
Off-farm Income	$22,626	$22,303	$27,935	$21,837
Total Cash Income	$78,654	$71,556	$108,386	$62,970
Farm Income/Acre	$59	$57	$75	$32
Total Farm Assets	$754,241	$766,761	$774,387	$468,130
Total Farm Debt	$128,917	$111,817	$166,545	$200,375
Farm Net Worth	$625,324	$654,944	$607,842	$267,755

10.3 The Design of Optimal Loan Contracts and the Role of Crop Insurance

In this section, we explore a theoretical model which shows that the typical lending contract structure is a rational response to information asymmetry. The model here follows the work of Gale and Helwig (1985). The loan contract implied by the model has several implications for crop insurance. Farmers with very low levels of debt or high cash incomes relative to their debt level will always be able to make required payments out of cash income; these farmers will be less likely to insure. Farmers with high levels of debt, but low net worth, have a limited liability—the model suggests that for such farmers, the crop alone will serve as collateral for the loan. These farmers have no self interest in crop insurance; however, the lenders have a strong interest in insurance; thus these farmers are more likely to be required to take out crop insurance. A third group of farmers have high debt, but significant net worth which can be pledged as collateral for the loan. These farmers will suffer a discrete cost (associated with liquidating capital) whenever income falls below the amount needed to make full repayment. In order to avoid this cost, such farmers are more likely to voluntarily insure.

The borrower starts with a known and fixed level of assets (A) and debt (D), on which payments are P(D). The borrower will borrow amount β (to be established by the contract), and will use that amount to purchase variable inputs and generate income I. Net Cash Farm Income is cash receipts from farm sources minus cash outlays for farm related expenses. Cash receipts include receipts from sales of crops and livestock, value of crops put under loan to the Commodity Credit Corporation, other income derived from government farm programs, value of farm output consumed by the farm family, the change in the value of inventories of crops and livestock during the year, income from custom work, grazing, insurance indemnity payments, and other farm related sources of income. Cash expenses include crop expenses such as seed, chemicals, fuel, and repairs, livestock expenses such as feed and feeder stock, non-family labor expenses, taxes, interest, utilities, insurance, and miscellaneous supplies.

Off-farm Income is calculated from a question in which farmers are asked into which range a certain source of income falls. For each source of off-farm income, the midpoint of the reported range serves as an estimate. These estimates are summed to give the total off-farm income figure reported here.

Total Farm Assets is an end-of-year figure for the market value of inventories, land, buildings, equipment, and financial assets (cash, accounts receivable, bank accounts) excluding "personal assets". Total Farm Debt is an end of year figure showing total outstanding principal plus unpaid interest on farm loans. Income will depend on assets, variable inputs, and a stochastic state of nature $\tilde{s}$: $I = I(A, \beta, \tilde{s})$.

The lender will acquire funds at a fixed per dollar cost of $(1 + i)$. The lender can observe everything about the borrower except realization of the state of nature s. The lender can observe s by incurring cost $C(s)$. The lender can be repaid either out of the borrower's cash income or out of the borrower's assets. For any amount κ of assets claimed by the lender in lieu of cash repayment, the lender will earn $L(\kappa)$. The (salvage) value of assets remaining in the hands of the borrower are $B(A - \kappa)$.

The contract between the borrower and lender specifies the amount borrowed (β), the states which will be observed by the lender (indicated by a function $\delta(s)$ which takes a value of 1 when the state s is observed, and a value of 0 when the state is not observed), the amount to be repaid by the borrower out of cash income when state s occurs ($\rho(s)$), and the amount of assets to be transferred by the borrower to the lender when state s occurs ($\kappa(s)$).

When state s occurs, the contract specifies that the income of the lender is $\rho(s) + L(\kappa(s)) - \kappa(s)C(S)$, and the income of the borrower is $I(\beta, A, s) - P(D) - \rho(s) + B(A - \kappa(s))$. The terms of the contract are constrained by the fact that the borrower's share of cash income and assets can never be negative: $\rho(s) \leq I(s) - P(D)$ and $\kappa(s) \leq A$. In addition, the terms of the contract are constrained to be "incentive compatible"—to ensure that the borrower accurately reports the state of nature which occurs (and on which the terms of the contract are based). Incentive compatibility is an extremely useful concept, since any optimal contract can be (weakly) dominated by an incentive compatible contract (see Harris and Townsend 1985).

In the present model, incentive compatibility is defined as follows:

Definition: A contract β, ρ, δ, κ is incentive compatible if and only if when state s occurs, then for any other state $\hat{s}$

(1) $\delta(\hat{s}) = 1$; or

(2) $\rho(\hat{s}) > I(s) - P(D)$; or

(3) $-\rho(s) + B(A - \kappa(s)) \geq -\rho(\hat{s}) + B(A - \kappa(\hat{s}))$.

If $\delta(\hat{s}) = 1$, the borrower cannot successfully misreport the true state as $\hat{s}$, since the lender will observe if the borrower reports $\hat{s}$ and will discover the true state is s. If $\rho(\hat{s}) > I(s) - P(D)$, then the borrower cannot misreport the true state as $\hat{s}$ because the borrower cannot pay the cash payment required when the state is $\hat{s}$. If $-\rho(s) + B(A - \kappa(s)) \geq -\rho(\hat{s}) + B(A - \kappa(\hat{s}))$, then the borrower would have no incentive to misreport the true state as $\hat{s}$ since borrower returns would be higher by reporting the true state. If at least one of these conditions holds for every false state $\hat{s}$, then the borrower will always report the true state s and the contract is incentive compatible.

In this paper, we assume that borrowers and lenders are risk neutral, and that lenders are competitive, so they are bound by a zero expected profits constraint. Thus the optimal contract problem can be written as:

$$\text{Max } E\ \{I - \rho - P + B(A - \kappa)\}$$

such that

(1) $E(\rho + L(\kappa) - \delta C - (1 + i)\lambda) = 0$

(2) $\rho \leq I - P \quad \kappa \leq A$

(3) the contract is incentive compatible.

The structure of the contract between the buyer and the lender will depend critically on the structure of costs of observing states and claiming borrower's capital. In this paper, we make the following assumptions:

(1) $C(s) = c$ for all s;

(2) $B(0) = L(0) = 0$; and $\lim_{\kappa \to 0} B(A - \kappa) = b \leq B(A)$, and $\lim_{\kappa \to 0} L(\kappa) = 1 < 0$;

(3) $\kappa_1 \geq \kappa_2$ implies $L(\kappa_1) + B(A - \kappa_1) \leq L(\kappa_2) + B(A - \kappa_2)$.

Assumption (1) states that observation costs of the lender are constant across states. This can be thought of as investigations and legal costs associated with obtaining a verified accounting of the borrower's income. Assumption (2) states that there are fixed costs associated with liquidating capital for the lender and that part of these fixed costs are borne by the lender. These fixed costs might be legal costs or other transactions costs associated with the sale and transfer of the capital assets. Assumption (3) states that there are varying degrees of fungibility of the borrower's assets. At very low levels of κ, the assets can be liquidated and transferred from borrower to lender with relatively low costs. As the level of κ rises, less liquid assets are sold off entailing larger and larger transactions costs.

These assumptions allow us to discover a general form for the optimal contract $\{\lambda^*, \rho^*, \kappa^*, \delta^*\}$ under which the following conditions must apply.

1. Capital assets will be claimed by the lender ($\kappa^* > 0$), only after all income has been claimed by the lender ($\rho^* = I - P$).

Proof. Suppose for some state $\tilde{s}$, $\rho^*(\tilde{s}) < I(\tilde{s}) - P$, and $\kappa^*(\tilde{s}) > 0$. Incentive compatibility requires either that $\delta(\tilde{s}) = 1$, or if $\delta(\tilde{s}) = 0$, for all states $s < \tilde{s}$ such that $I(s) - P > \rho^*(\tilde{s})$, that $-\rho^*(s) + B(A - \kappa^*(s)) \leq \rho^*(\tilde{s}) + B(A - \kappa(\tilde{s}))$. In either case, there exists a contract $\{\lambda^*, \hat{\rho}, \hat{\kappa}, \delta^*\}$ with $\hat{\rho} = \rho^*$ and $\hat{\kappa} = \kappa^*$ in states

$s \neq \tilde{s}$, $\hat{\rho}(\tilde{s}) = I(\tilde{s}) - P$, $\hat{\kappa}(\tilde{s}) \leq k^*(\tilde{s})$, such that $-\rho^*(\tilde{s}) + B(A - \kappa^*(\tilde{s})) = \hat{\rho}(\tilde{s}) + B(A - \hat{\kappa}(\tilde{s}))$. The contract $\{\lambda^*, \hat{\rho}, \hat{\kappa}, \delta^*\}$ is incentive compatible and dominates $\{\lambda^*, \rho^*, \kappa^*, \delta^*\}$ since borrower returns are identical under both contracts and lender returns are identical in all states except state $\tilde{s}$ where they are higher under contract $\{\lambda^*, \hat{\rho}, \hat{\kappa}, \delta^*\}$: $\rho^*(\tilde{s}) - \hat{\rho}(\tilde{s}) = B(A - \kappa^*(\tilde{s})) - B(A - \hat{\kappa}(\tilde{s})) \leq L(\hat{\kappa}(\tilde{s})) - L(\kappa^*(\tilde{s}))$. (The equality is by construction and the inequality by cost condition (iii).) Thus, $\rho^*(\tilde{s}) + L(\kappa^*(\tilde{s})) \leq \hat{\rho}(\tilde{s}) + L(\hat{\kappa}(\tilde{s}))$.

2. There exists a level of capital $\underline{\kappa}$ such that for all $\kappa < \underline{\kappa}$, $L(\kappa) \leq 0$. If the state s is observed, $\kappa^*(s) > \underline{\kappa}$.

3. Depending on the level of observation costs c and fixed costs of claiming capital I, there may exist a level of capital $\underline{\underline{\kappa}} < \underline{\kappa}$ such that for all $\kappa < \underline{\underline{\kappa}}$, $L(\kappa) \leq -c$. For states in which $\delta = 0$, $\kappa^*(s) > \underline{\underline{\kappa}}$.

Otherwise $\delta(s) = 1$ and $\hat{\kappa}(s) = 0$ is incentive compatible and gives higher returns to both borrower and lender.

4. Suppose $\kappa^*(\tilde{s}) > 0$ for some $\tilde{s}$. Then for all $s < \tilde{s}$, either $\kappa^*(s) > \kappa^*(\tilde{s})$, or $\delta^*(s) = 1$. This follows from incentive compatibility and result 1.

5. Suppose $\kappa^*(\tilde{s}) \geq 0$, $\delta^*(\tilde{s}) = 0$ for some $\tilde{s}$. Then for all $s > \tilde{s}$, $\kappa^*(s) \geq \kappa^*(\tilde{s})$.

Results 1, 3, and 4 show that capital assets can be used to enforce incentive compatibility only over a range of states limited by the availability of capital. The amount of capital claimed in any state must be in the range $(\underline{\underline{\kappa}}, A)$. Incentive compatibility requires that the range of gross income I in states in which capital is used to enforce incentive compatibility cannot exceed $B(A - \underline{\kappa})$. Outside this range of states, incentive compatibility must be enforced by direct observation, or by setting ρ equal to a constant, and setting $\kappa^* = 0$ (by result 1).

These results suggest that an optimal debt contract will take the following general form:

1. For $s \geq \bar{s}$, $\delta^*(s) = 0$, $\rho^*(s) = I(\bar{s}) - P$, and $\kappa^*(s) = 0$.

2. For $\underline{s} < s < \bar{s}$, $\delta^*(s) = 0$, $\rho^*(s) = I(s) - P$ and $\kappa^*(s) > 0$.

3. For $s < \underline{s}$, $\delta^*(s) = 1$.

More common terminology can be used to describe this contract as follows. The range $s \geq \bar{s}$ describes income levels in which cash income is sufficient to make "full repayment". Incentive compatibility is enforced in this range by having the same repayment terms for all states. The range $\underline{s} < s < \bar{s}$ describes a range of income in which the farm is "illiquid" (insufficient cash to make full repayment). In this range, incentive compatibility is enforced in two ways. The farmer is unable to report a higher state than the true state, because actual income is insufficient to make the repayment required under a higher state. The farmer is unwilling to report a lower (unobserved) state than the true state because costs to the farmer are higher when lower (unobserved) states are reported. The range, $s \leq \underline{s}$, describes a range of incomes in which "bankruptcy" occurs. Here, the total assets of the farmer are observed by the lender (for example, through a bankruptcy proceeding). Observation precludes misreporting of states.

For a farmer with sufficiently low level of debt, or sufficiently high income, all of the probability weight is assigned to states $s \geq \bar{s}$. These farmers have zero probability of illiquidity or bankruptcy. Clearly, these farmers have no financial incentive to insure crops. As Fig. 1(a) shows, for farmers in this situation, as cash income increases by one dollar, net borrower profits increase by one dollar; the debt, if any, of these farms poses a lump sum tax on the borrower, which is not influenced by the level of income.

As noted above, the amount of capital claimed by the lender will always be in the range $(\underline{\kappa}, A)$. For farms with sufficiently low

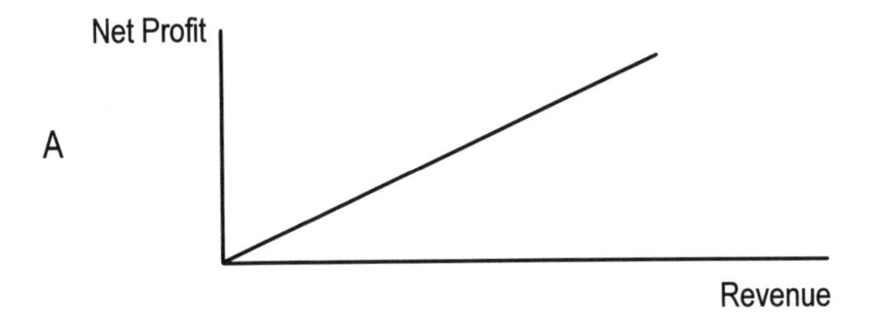

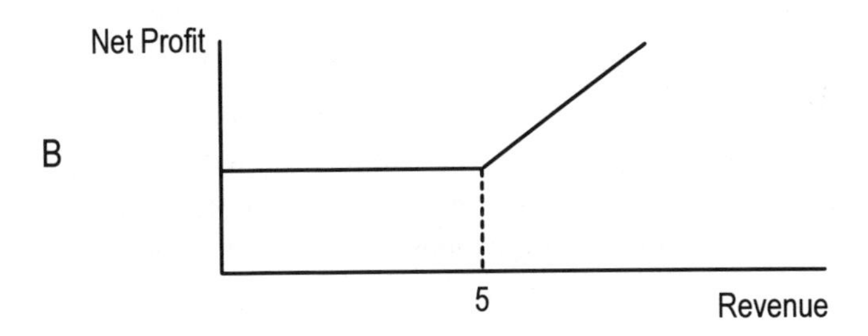

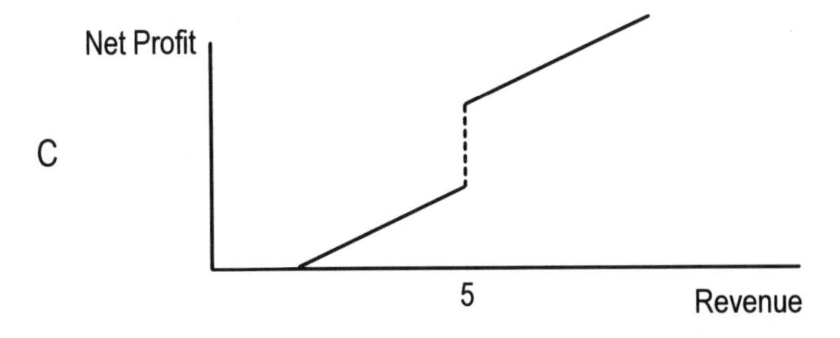

Fig. 1. Net Returns to borrower as a function of cash revenues, under various loan contracts

A and sufficiently high $\underline{\underline{\kappa}}$, this space may be empty. The value of maximum amount of capital which is available for claiming by the lender is insufficient to cover the lender's fixed costs of claiming capital. Thus, for these farms, the lender will never claim borrower's capital, and capital plays no role in the loan contract.

In this case, as shown by Gale and Helwig (1985), the optimal contract is one in which the borrower has limited liability. In states above some $\overline{s}$, borrowers will repay a fixed amount, and lenders will not observe the state. In states below $\overline{s}$, lenders will observe the state and claim all income.

The borrower's net income mapped against cash revenues is shown in Fig. 1(b). For revenues above those associated with state $\overline{s}$, the net profit function is the same as for the low debt farmer shown in Fig. 1(a). But when low revenues occur, the borrower turns over all revenues to the lender, and keeps all capital. Thus in all states below $\overline{s}$, the borrower's profit is the same. The effect of this contract is to make the borrower's profit function convex over cash revenue. The (risk neutral) borrower will not be interested in crop insurance even if the insurance is perfectly fairly priced. On the other hand, the lender's profit function is now concave over cash revenue, and the (risk neutral) lender does have an interest in crop insurance. For farmers in this situation, there is a clear financial incentive for the lender to require the borrower to take out crop insurance.

Finally, we address the question: Are there financial circumstances in which a risk neutral borrower would voluntarily take out fair crop insurance? The answer to this yes, if there are states in which capital claimed by the lender is used to enforce incentive compatibility. As shown above, in such states, the net profits to the borrower must be lower in lower states; otherwise, the borrower has an incentive and the ability to misrepresent the true state and report instead some lower unobserved state. If in addition there are positive fixed costs of liquidating capital to the lender and borrower ($l > 0$ and $b > 0$), $\underline{\underline{\kappa}} > 0$, and there will be a discrete jump in the borrower's net profit function at the state $\overline{s}$. Thus, we expect the net profits function to have the shape shown in Fig. 1(c). Here the incentive for the borrower to insure crops is to reduce the probability of revenues to the left of that discrete jump.

10.4 Econometric Evidence on the Role of Financial Characteristics in the Decision to Insure Crops

In the last section some testable hypotheses concerning crop insurance and financial characteristics of the firm were derived: farmers with low debt and high incomes are less likely to insure; farmers with high debt and low net worth are more likely to be required to insure; and farmers with high debt and high net worth are more likely to insure voluntarily. These hypotheses are tested in this section, using the Crop Insurance sample of the Farm Crop and Resource Survey described above. The dependent variables take values of zero or one, making Logit (or Probit) regression techniques appropriate. To take account of the sample selection, weights were applied to the raw data before regression, utilizing the SAS program PROC LOGIST (see Harrell 1986 for a description).

The regression results testing the first hypothesis are reported in Table 3. The dependent variable in these regressions takes a value of 1 when the farm did not purchase crop insurance, and a value of 0 when the farm did purchase crop insurance. Our hypothesis says that this variable should be positively related to income and negatively related to debt. Regression 1 shows that the decision not to insure is significantly affected by level of debt. As expected, the coefficient on farm debt was negative and significant. Cash income was expected to be positively correlated with the decision to not insure; thus the coefficient on cash income in regression 1, though insignificant, is the "wrong" sign. There are at least two possible explanations for this. First, since crop insurance pays out on a per acre basis, it may be income per acre which is the appropriate measure in this context. As regression 2 shows, when cash income is expressed on a per acre basis, the coefficient is positive and significant. The negative and significant coefficient on debt is not affected by this change. A second possibility is that the appropriate factor influencing a decision not to insure is the degree to which income covers debt. Farmers with a high debt-coverage ratio would be expected to be more likely to not insure. Regression 3 shows that there is a significant positive

Table 3. LOGIT regression results, financial characteristics, and the decision to not insure based on the crop insurance subsample of Farm Crop and Resource Survey, 1988

$$\text{No Insurance} = \underset{(0.065)}{1.347} - \underset{(5 \times 10^{-8})}{8 \times 10^{-8}} \cdot \text{Total Cash Income} - \underset{(1 \times 10^{-10})}{2 \times 10^{-9}} \cdot \text{Farm Debt} \quad (1)$$

$$\text{No Insurance} = \underset{(0.073)}{1.075} + \underset{(0.0003)}{0.0013} \cdot \text{Total Cash Inc/Acre} - \underset{(1 \times 10^{-10})}{2 \times 10^{-9}} \cdot \text{Farm Debt} \quad (2)$$

$$\text{No Insurance} = \underset{(0.063)}{0.929} + \underset{(1 \times 10^{-9})}{7 \times 10^{-9}} \cdot \text{Debt Coverage Ratio} \quad (3)$$

Standard errors are reported in parentheses.

All coefficients are significant at the 95% level except for the coefficient on Total Cash Income in equation 1.

Debt Coverage Ratio = Net Cash Farm Income/Farm Debt

The "No Insurance" variable takes a value of 1 when the farm did not buy crop insurance, and takes a value of 0 otherwise.

relationship between debt coverage and the decision not to insure.

Regression results testing the second hypothesis are shown in Table 4. The dependent variable in these regressions takes a value of 1 when the farm was required to insure by a lender and takes a value of 0 when the farm insured voluntarily. Farms who did not purchase crop insurance are not included in this analysis. The reason for excluding these farms is that while our theory clearly differentiates between those who have high debt and a large amount of claimable assets (voluntary insurers) and those who have high debt and low amount of claimable assets (required insurers), it does not have prediction whether the claimable assets of the noninsurers will be high or low compared to noninsurers. A farm with low claimable assets and low debt would not insure, and a farm with low claimable assets and high debt would be forced to insure. Our hypothesis here is that the dependent variable should be negatively related to a farm's net worth.

In spite of the apparently large differences in average net worth between the two groups, it was not possible with the SAS

Table 4. LOGIT regression results: financial characteristics and the decision to require insurance based on the crop insurance subsample of Farm Crop and Resource Survey, 1988

$$\text{Required} = \underset{(0.166)}{-1.321} - \underset{(1.4 \times 10^{-6})}{1.0 \times 10^{-6}} \cdot \text{Tot Cash Inc} - \underset{(\text{N.R.})}{0.001} \cdot \text{Net Worth} \qquad (4)$$

$$\text{Required} = \underset{(.692)}{-0.273} - \underset{(1.3 \times 10^{-7})}{1.2 \times 10^{-7}} \cdot \text{Tot Cash Inc} - \underset{(0.059)}{0.112}\ \text{W ln (Net Worth)} \qquad (5)$$

$$\text{Required} = \underset{(0.150)}{-1.800} + \underset{(0.213)}{0.414}\ \text{W Debt/Asset} \qquad (6)$$

Standard errors are reported in parentheses.

All coefficients are significant at the 95% level except for the intercept and the coefficient on Total Cash Income in equation 5, and the coefficient on Total Cash Income in equation 4. The standard error for the coefficient on Net Worth in equation 4 was not reported (N.R.), because the variable had limited dispersion. See the text for a fuller explanation.

The "Required" variable takes a value of 1 when the farm was required to buy crop insurance, and takes a value of 0 when the farm bought insurance voluntarily.

LOGIST software to obtain a straightforward statistical test of this hypothesis. In every attempt, the software detected "limited dispersion" of the dependent variable Net Worth and reported a coefficient which was (invariably) negative (the expected sign) but reported no standard error for the estimated coefficient, making hypothesis testing impossible. The results of such a regression are presented as regression 4. Attempts to trick the computer by rescaling the variable (dividing the numbers by up to 1,000,000, or multiplying by 100) or changing the zero-one orientation of the dependent variable were unsuccessful. Rescaling the net worth variable using natural logs required the dropping of 23 observations of farms with negative net worth, since log of a negative number is not defined. This rescaling was successful in getting results, shown in regression 5. Here the coefficient on the net worth variable is negative and significant as expected. Regression 6 uses the Debt/Asset ratio rather than Net Worth (Assets - Debt). Here, as expected, the coefficient was positive and significant.

Further work is needed on the relation of net worth to the decision to require insurance. Nevertheless, the results presented here provide some solid support for the theory that financial characteristics of the firm play a role in influencing decisions concerning crop insurance.

10.5 Implications for Crop Insurance Analysis and Policy

The results of this paper have implications both for the analysis of crop insurance participation and for the government policy affecting crop insurance decisions.

The most obvious implication is that models attempting to explain the participation decision should include debt as an explanatory variable. Farms with low levels of debt have low (zero) probabilities of suffering from illiquidity; neither lenders nor borrowers have an incentive to take out crop insurance in order to avoid costs associated with illiquidity.

A second implication for modelling participation is that insurers can be divided into two distinct groups: those who insure voluntarily and those who are required to insure. The 20 percent of insurers who are involuntary insurers make up a sizable portion of total insurers. On average, the characteristics of the two groups differ substantially in a number of areas (see Table 2). The possibility exists that analyses of participation which bifurcate the whole population into insurers and non-insurers will obtain unclear results because of the unwarranted union of voluntary and involuntary insurers into a single group. For example, voluntary insurers seem to have much larger farms than non-insurers (2086 acres compared to 1228), but the difference between noninsurers and involuntary insurers is much smaller. Even more dramatically, farm income per acre is larger (on average) for voluntary insurers than for non-insurers, but is smaller for involuntary insurers than for non-insurers. The difference between farm income per acre for insurers and for noninsurers appears to be quite small. These differences emphasize the need for a clear theoretical base for empirical analysis of insurance participation decisions. If the theory models voluntary decisions to insure, then empirical tests of that theory will be more appropriate if they exclude involuntary insurers from the sample.

What lessons can policy makers draw from this paper? Perhaps the most important question raised by the above analysis concerns the economic purpose of crop insurance. The results suggest that insurance may reduce the incidence and aggregate size of transactions costs associated with illiquidity and/or insolvency. It remains to be determined whether this role is incidental to the risk management role of insurance. However, if financial reasons for insuring are paramount, there are a significant number of farms (farms with low levels of debt) for whom crop insurance may never be an important management tool.

This observation leads to several other conclusions. First, high participation in crop insurance should not be an end in itself. The focus should remain on crop insurance as a means of accomplishing other objectives. Second, the fact that many farmers choose not to insure is not necessarily evidence of inherent imperfections in the crop insurance program. The model presented here shows that low debt farmers will have no interest even in a "perfect" insurance plan. Third, proposals to increase participation in insurance by having all FmHA and FCS loans require insurance are probably misguided. Requiring insurance of some borrowers (those with high debt and low net worth) is a rational response. But the requirement should be imposed on a case by case basis.

10.6 Conclusions

This paper explored the possibility that crop insurance decisions were influenced by financial characteristics of the farm. We found that a not insubstantial proportion (20 percent) of all participants in crop insurance participate as a condition of a loan. Further, there appear to be differences in the characteristics of voluntary and involuntary insurers.

The theory presented in section 10.2 shows how bankruptcy and liquidation costs influence the structure of the loan contract. The theory suggests that farmers with low debt are more likely to choose not to insure, and that those with high debt and low net worth are more likely to be required to insure. These hypotheses are supported by the econometric analysis of the data from the Farm Crop and Resource Survey crop insurance subsample.

This evidence supports the general notion that crop insurance plays a role in reducing transactions costs. In this regard, insurance may provide a benefit to the borrower, to the lender, to both, or to neither. In analyzing crop insurance participation it may be misleading to lump together voluntary insurers with involuntary insurers. In developing policy options, it is a mistake to assume that all farms would reap a (transactions costs reduction) benefit from insurance.

Acknowledgment. Many thanks to Agapi Somwaru, Linda Calvin, and Joe Glauber of USDA, ERS, without whom the empirical section of this paper would have been impossible.

References

Calvin L (1990) Participation in federal crop insurance. Paper presented at S Agr Econ Assn, Little Rock, Arkansas. USDA, ERS, Washington, DC

Commission for the improvement of the federal crop insurance program (1989) Recommendations and findings to improve the federal crop insurance program. Washington, DC

Fuller W (1975) Regression analysis for sample survey. Indian J Stats 37:117-132

Fuller W (1984) Least squares and related analyses for complex survey designs. Survey Methodology 10:97-118

Gale D, Helwig H (1985) Incentive compatible debt contracts: the one-period problem. Rev Econ Studies 52:647-664

Glauber JW, Harwood JL, Miranda MJ (1989) Federal crop insurance and the 1990 farm bill: an assessment of program options. USDA, ERS, CED, staff report no. AGES 89-45, Washington, DC

Harrell FE Jr (1986) The LOGIST procedure. Chapter 23 in SAS Users Group, Int., Supplementary user's guide, version 5 edition. SAS Institute Inc., Cary, North Carolina

Harris M, Townsend R (1985) Allocation mechanisms, asymmetric information and the revelation principle. In: Feiwel G (ed) Issues in contemporary microeconomics and welfare. SUNY Press, Albany, New York

Kaplan B, Francis I (1979) A comparison of methods and programs for computing variances of estimators from complex sample surveys. Proceedings of the Section on Survey Research Methods, American Statistical Association

Leatham DJ, McCarl BA, Richardson JW (1987) Implications of crop insurance for farmers and lenders. Southern J Agr Econ 19:113-120

Pflueger BW, Barry PJ (1986) Crop insurance and credit: a farm level simulation analysis. Agr Finance Rev 45:1-14

Chapter 11

Risk Reduction from Diversification and Crop Insurance in Saskatchewan

R.A. SCHONEY, J.S. TAYLOR[1] and K. HAYWARD[2]

11.1 Introduction

Most farm production decisions are made in a stochastic and dynamic environment. Major stochastic variables include prices, crop yields, interest rates, and government farm policy. Agricultural production is also characterized by dynamic processes: many biological processes unfold over time and are linked to previous decisions through soil moisture, soil fertility and crop disease. Farm financial relationships are also dynamic: farm profits or losses impact on cash flows, and farm cash surpluses/deficits impact on future cash flows through decreased/increased debt service or increased/decreased investments. In a dynamic and stochastic environment, risk attitudes and risk bearing ability are also important variables influencing crop production. Yet many governmental policies and extension recommendations are based on profit maximizing rotations which ignore risk. However, according to Wilson and Eidman (1983), fundamental decisions such as crop selection cannot be isolated from risk management and dynamic considerations. In Saskatchewan, three decisions must be made more or less simultaneously: crop selection, cropping intensity and input intensity. While the latter two are important, only crop

---

[1]Department of Agricultural Economics, University of Saskatchewan, Saskatoon, Saskatchewan, Canada.

[2]Saskatchewan Crop Insurance Corporation.

selection under risk is treated here.[3] This paper examines crop selection under risk using the crop portfolio model, surveys the empirical relationships between alternative crop candidates on the black soils of Saskatchewan, and contrasts the risk reduction between diversification and crop insurance.

11.2 Crop Portfolios

Cropping decisions in Saskatchewan can be examined using a crop portfolio whose rudiments were laid out by Anderson, Dillon and Hardaker in 1977. In modern crop portfolio analysis, farm cropping decisions are considered in a fashion similar to a stock portfolio. The proportion of land devoted to fallow or crop is similar to the decision to invest in a particular stock, except that fallowing/cropping decisions are also dynamic. In crop portfolio analysis, the decision maker's objective is to maximize expected returns subject to a predetermined risk level. By parametrically varying the risk level, the efficient income-risk frontier can be mapped and the decision maker chooses an optimum portfolio based on his/her individual risk preferences.[4] Key to the analysis is the relationship between crop portfolio returns and risk. The expected return to the crop portfolio is

$$E(G) = \sum_{c=1}^{n} [E(P_c)\, E(Y_c) + \mathrm{Cov}\,(P_c, Y_c)]\; A_c \tag{1}$$

where E = expectations operator
c = crop number,
n = number of crops,

---

[3]Input intensity associated with fallow crops is already low and probably not very price or risk responsive. The input intensity associated with stubble cropping is far more responsive.

[4]Note that the existence of an actuarially sound insurance may alter the shape of the portfolio (Newberry and Stiglitz, 1981).

G = gross farm income,
P = farmgate commodity price,
Y = yield per acre,
A = harvested acreage, and
Cov(P,Y) is the covariance of price and yield.

For most crops, the farm level covariance between price and yield is zero.[5]

Key to the crop portfolio approach is the effect of differing crop combinations on gross income variability. An approximation of variance of gross returns derived by Anderson et al. (p. 33, 1977) is given by:

$$\mathrm{Var}(G) \approx \sum_{i=1}^{n} [E(P_i)^2\,\mathrm{Var}(Y_i)\,A_i^2] + \mathrm{Var}(P_i)^2\,E(Y_i)^2\,A_i^2] \tag{2}$$

$$+ 2\sum_{i=1}^{n}\sum_{j=1}^{n} [E(P_i)\,E(P_j)\,E(Y_i)\,E(Y_j)\,A_i\,A_j\,\mathrm{Cov}\,(P_i, Y_j)]$$

$$+ 2\sum_{i=1}^{n}\sum_{j=1}^{n} [E(P_i\,E(P_j)\,E(Y_i)\,E(Y_j)\,A_i\,A_j\,\mathrm{Cov}\,(P_i, P_j)]$$

$$+ 2\sum_{i=1}^{n}\sum_{j=1}^{n} [E(P_i)\,E(P_j)\,E(Y_i)\,E(Y_j)\,A_i\,A_j\,\mathrm{Cov}\,(Y_i, Y_j)]$$

where E = expectations operator and
Cov (P_i, P_j) = covariance of price_i and price_j,
Cov (Y_i, Y_j) = covariance of yield_i and yield_j and
Cov (Y_i, P_j) = covariance of yield_i and price_j.

---

[5]This may not be the case with all crops, however. Specialty crops with thin markets and grown in relatively concentrated regions may have strongly correlated farm yields and area yields. If the area faces a downward sloping demand curve, then farm commodity prices are indirectly and negatively correlated with farm yields. For a further discussion refer to Weisensel and Schoney (1989).

From equation (2), gross income variability is a complex function of yield and price variability and the covariance between all prices, all yields and all price-yield combinations. For most crops, the amount of potential risk reduction that can be achieved by diversifying the crop portfolio is determined by the degree of association between the various commodity yields and between the various commodity prices. In the following analysis, it is convenient to use yield-yield correlations, τ_{yy} and price-price correlations, τ_{pp}, instead of the covariances. It is obvious that low τ_{yy} and τ_{pp} will lead to relative uncorrelated gross returns and thus, enhanced diversification potential. But for an individual farm, crop yields are contemporaneously related through weather patterns, leading to high yield-yield correlations between closely related crops such as cereals. For example, yield-yield correlations, τ_{yy}, are typically quite high for most dryland cereal crops, usually in the range of 0.8 to 0.9. However, the τ_{yy} between less closely related crops is somewhat lower; in the case of cereal, oil seed or specialty crops the τ_{yy} lower are in the range of 0.6 to 0.75 (Table 1). Likewise, commodity prices may be also be contemporaneously related through inter-related commodity markets. Hence, the price-price correlations, τ_{pp}, between cereals is very high, 0.95 but the τ_{pp} between lentil and wheat prices is quite low (0.26). The combined effect of the price-price and yield-yield correlations can be expressed as τ_{ii}, the correlation between gross income of alternative crops. Because of both the high τ_{pp} and τ_{yy}, most Saskatchewan crops have relatively high τ_{ii}. For example, using wheat-barley, the τ_{pp} is 0.95, the τ_{yy} is 0.91, and the resulting τ_{ii} is 0.88. Alternatively, barley-lentil is a more attractive diversification combination, because the τ_{pp} is 0.24, the τ_{yy} of is 0.62 and the corresponding τ_{ii} is 0.02.

Table 1. Cross correlations of yield, price and gross income, by crop, Saskatchewan

Interaction/ Crop	Crop				
	Wheat	Barley	Canola	Flax	Lentil
Price-Price:					
Wheat	1.00	0.95	0.80	0.92	0.26
Barley		1.00	0.81	0.90	0.24
Canola			1.00	0.86	0.36
Flax				1.00	0.24
Lentils					1.00
Yield-Yield:					
Wheat	1.00	0.91	0.64	0.85	0.72
Barley		1.00	0.73	0.93	0.62
Canola			1.00	0.85	0.76
Flax				1.00	0.78
Lentils					1.00
Gross Income:					
Wheat	1.00	0.88	0.70	0.91	0.05
Barley		1.00	0.52	0.86	0.02
Canola			1.00	0.76	0.44
Flax				1.00	0.04
Lentils					1.00

Source: Yields: Saskatchewan Crop Insurance Data, 1958-1988
Prices: Agricultural Statistics, 1989
Gross Income correlations are calculated.

11.3 Gross Margin Correlations and Diversification

In order to demonstrate the effect of τ_{yy} and the relationship between the size of the portfolio and risk, the following example is presented. Assume that a farmer can form a crop portfolio from n identical crops, each having an expected gross margin of 20% crop and a standard deviation of 5%. Also, assume returns are normally distributed, all crops are held in equal proportions and that the gross margin correlations between all crops are either 0.2, 0.5 or 0.8. Finally, the portfolio risk is measured by the coefficient of variation (CV), defined as the standard deviation divided by return. The effect of diversification on the portfolio CV as additional crops are added to the crop portfolio, is shown in Fig. 1. When crop gross margins are very highly correlated, as in this example a correlation of 0.8, there is little benefit from diversifying the crop portfolio; what little benefits that can be derived are exhausted by moving from a one crop to two crop portfolio. Even when the gross margin correlations are reduced to 0.5, most of the benefits of diversification are still achieved by moving from a one crop to a two crop portfolio; the relative riskiness of a one crop portfolio is 25.0% while the relative riskiness of a two crop portfolio is 21.7%, a difference of 3.3%. The relative riskiness of

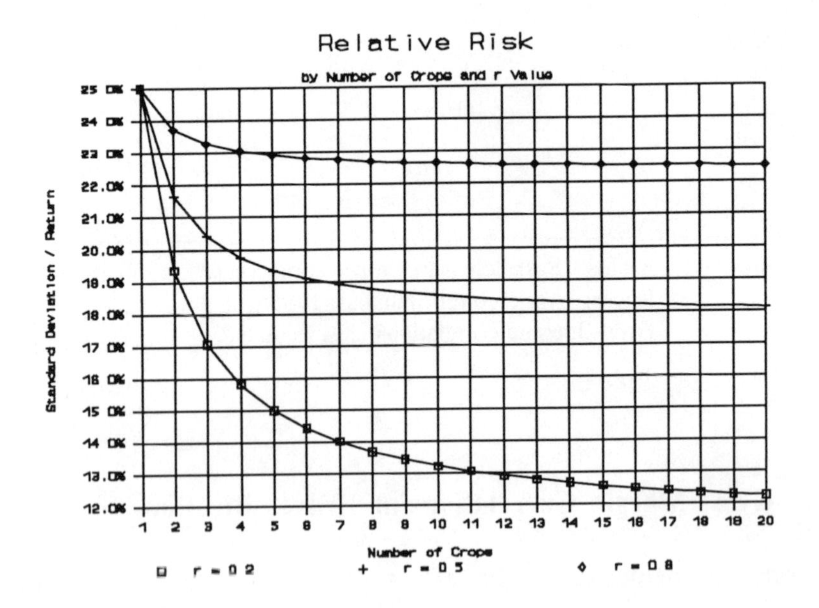

Fig. 1. The impact of adding crops with correlated returns on portfolio risk, correlation coefficient = 0.2,0.5 and 0.8

a three crop portfolio is 20.4%, which is only 1.2% less than the two crop portfolio. Most of the benefits of diversification are derived with a relatively small portfolio because of the rapidly diminishing benefits of adding more crops.[6] When the correlation coefficient is reduced to 0.2, then the addition of alternative crops reduces risk substantially more but the effect still rapidly diminishes. In this case a producer might choose 3 or 4 crops as compared to the two crops in the previous cases.

Since most Saskatchewan crops such as wheat, barley and canola all have a τ_{ii} of 0.5 or higher, the conclusions for Saskatchewan are obvious. Even if all crops were equally profitable, there is little potential for risk reduction in diversifying the crop portfolio beyond two or three crops.

11.4 A Crop Portfolio on the Black Soils of Saskatchewan

Not all crops, however, are equally profitable and yield-yield correlations are somewhat more complex in that stubble crops are less highly correlated with fallow crops than with other stubble crops. Likewise, farmers can choose from a variety of crop insurance options. In order to demonstrate the impact of crop diversification on whole farm risk, a portfolio of four commonly grown crops is assembled for a representative grain and oilseeds farm in the black soil zone of Saskatchewan. Four crop portfolios are defined: wheat, wheat-canola, wheat-canola-barley and wheat-canola-barley-oats (Table 2). All crop portfolios maintain 25% of land in summerfallow. The representative farm has a total of 1740 acres. The various cost and return relationships are based on published costs and returns from the 1991 Top Management Workshops (Saskatchewan Agriculture and Food).

Net cash available is simulated 200 times for each of the four crop portfolios under the no crop insurance and crop insurance

---

[6]This results from the assumption of constant correlation. Thus, as additional crops are added, the number of covariances rises rapidly.

Table 2. Assumed crop portfolio composition, by number of crops

Crop	Crop Portfolio Size (Number of Crops)			
	ONE	TWO	THREE	FOUR
	(acres)			
WHEAT ON FALLOW	435			
WHEAT ON STUBBLE	870	1,305	435	435
CANOLA ON FALLOW		435	435	435
BARLEY ON STUBBLE			435	217
OATS ON STUBBLE				218
FALLOW	435	435	435	435
TOTAL	1,740	1,740	1,740	1,740

alternatives using the Top Management Farm Business Simulator (TMFBS) (Schoney, 1991). The resulting cumulative distribution of cash is presented in Table 3 for varying size of crop portfolio for both the no crop insurance and crop insurance alternatives. The TMFBS takes into account the various production, cost and financial linkages as well as the various stochastic relationships. Net cash available is the amount of cash available to the farm family after income taxes and basic family living withdrawals have been deducted from cash flows. This amount represents discretionary income in that it is the amount available for capital replacement, new capital investments or increased family consumption. While a farm family can survive occasional modest cash flow deficits by increasing operating capital loans, chronic cash flow deficits leads to financial stress, net worth erosion and ultimately to farm business failure. Finally, the set of risk efficient

Table 3. Simulated cumulative distribution of net cash available, by crop portfolio and insurance alternative

Percentile	No Crop Insurance				Crop Insurance			
	Crop Portfolio				Crop Portfolio			
	One	Two	Three	Four	One	Two	Three	Four
Minimum	($86,115)	($38,768)	($67,377)	($56,980)	($62,877)	($28,265)	($46,778)	($47,560)
1st Percentile	($65,548)	($29,632)	($48,251)	($49,744)	($46,529)	($19,921)	($41,931)	($40,535)
5th Percentile	($46,177)	($12,158)	($33,535)	($29,888)	($39,033)	($8,508)	($25,900)	($24,124)
10th Percentile	($38,592)	$896	($25,612)	($19,840)	($33,062)	$1,497	($20,285)	($16,608)
20th Percentile	(25,705)	$8,711	($9,468)	($10,622)	($22,056)	$8,048	($8,321)	($8,373)
30th Percentile	($14,417)	$13,890	($2,674)	($3,757)	($15,364)	$14,158	($449)	($1,209)
40th Percentile	($6,538)	$19,116	$4,479	$7,242	($8,028)	$17,466	$5,992	$5,985
50th Percentile	$1,091	$22,863	$11,005	$12,079	($1,207)	$21,799	$10,431	$11,736
60th Percentile	$7,856	$27,237	$15,240	$15,729	$5,197	$25,983	$15,431	$15,003
70th Percentile	$15,290	$32,748	$21,327	$19,187	$13,405	$32,015	$19,999	$18,363
80th Percentile	$23,534	$41,897	$26,325	$26,656	$21,572	$39,646	$25,703	$25,321
90th Percentile	$32,876	$48,381	$36,319	$34,893	$30,982	$46,382	$35,539	$35,347
95th Percentile	$41,346	$58,985	$44,197	$39,763	$39,452	$56,242	$41,378	$38,720
99th Percentile	$57,667	$70,839	$55,786	$55,862	$55,915	$68,096	$53,166	$53,261
Maximum	$66,914	$82,914	$69,348	$65,359	$65,152	$80,171	$66,728	$62,809
Mean	($782)	$23,692	$7,934	$8,562	($515)	$23,011	$9,100	$9,091
Net Payout[a]					$267	($681)	$1,166	$529
Standard Deviation	$27,835	$20,644	$23,860	$22,173	$24,624	$18,948	$20,969	$20,138
Worst Event Protection[b]					$23,238	$10,503	$20,599	$9,420

[a]The Net Payout is defined as the mean of the insured portfolio less the mean of corresponding mean of the uninsured portfolio.

[b]The Worst Event Protection is defined as the minimum value of the insured portfolio less the mean of corresponding minimum value of the uninsured portfolio.

portfolios is selected using the familiar first and second degree stochastic dominance techniques.

11.5 Risk Efficient Crop Portfolios Without Crop Insurance

One of the major problems in diversification in Saskatchewan is that there are not very many profitable crops. Of the three crops examined, canola is the most profitable but it is limited to one year in four by disease restrictions. Wheat is more profitable than either barley or oats, and barley and oats are approximately equally profitable. Thus, a shift from a one-crop (wheat) to a two-crop (wheat-canola) portfolio, results in dramatically increased net cash available due to the greater canola profitability. While more diverse portfolios result in less variability, they are also considerably less profitable. Hence, the two-crop (wheat-canola) crop portfolio is the first degree stochastically dominant portfolio and would be preferred by all farmers who possess positive marginal utility.

11.6 Risk Efficient Crop Portfolios and Crop Insurance

In addition to crop diversification, farmers can enroll in the Saskatchewan Federal-Provincial all-risk crop insurance program to reduce risk. While crop diversification can provide both yield and price risk relief, crop insurance is primarily designed to reduce yield variability. In the 1990/91 crop-year, there were two yield options: 60% and 70% of expected yield, respectively and two price options: the base price (which was roughly the initial Wheat Board price where applicable) and 80% of the base. In order to demonstrate the combined effect of crop insurance and crop portfolios, the various crop portfolios are combined with the 60% yield protection and base price plan and resimulated using the same underlying yield and price observations generated in the previous scenario. The results are displayed in Table 3. Because

Saskatchewan crop insurance premiums are subsidized by Federal and Provincial governments, in the aggregate, payouts should exceed premiums paid by farmer. However, in actual practice, the amount of subsidy varies by crop, soil zone and farmer. This is shown as the net payout, the difference between the uninsured mean cash and the corresponding insured mean cash in Table 3. The simulations show a program that is relatively neutral—most crops such as wheat, oats and barley have a slight net positive payout but canola has a small net negative payout or a net premium.[7] For example, using the dominant and prevailing wheat-canola rotation, premiums exceed payouts by an average of $681 per year. The result is that the two crop, insured portfolio cannot dominate the uninsured, two crop portfolio. While the other crops and cropping rotations have small subsidies, the two crop portfolio remains stochastically dominant.

Given the relatively high cropping intensity and input costs of the black soil zone, crop insurance primarily acts as insurance against near catastrophic events: payouts average approximately one year in eleven. In order to show relative amount of catastrophic insurance, the worst event protection is calculated. The worst event protection is based on the difference in the minimum value of each portfolio scenario or in this case, the worst event in 200 simulations. Again, using the dominant two crop rotation, the minimum net cash available without crop insurance is -$38,768; with crop insurance, the minimum net cash available is -$28,265, or a worst event protection of $10,503. Note, however, the worst event protection varies by portfolio, with most protection of $23,238 being provided for the one-crop portfolio.[8] Hence, the diversifying effect of the portfolio offsets some of the protection offered by crop insurance.

---

[7]Of course, the results will be sensitive to the underlying statistical parameters. This is particularly true of canola which is a comparatively recent crop.

[8]Some of the differences may be due to the randomness associated with the stochastic simulation process.

11.7 Summary and Conclusions

Crop selection is considered within the context of a crop portfolio: the farmer selects a portfolio of crops (including fallow/stubble crops). Important considerations in portfolio selection include not only individual mean-variance but crop income correlations. However, the gross incomes of most common Saskatchewan crops are highly correlated, indicating that even if all crops were equally profitable and equally risky, there is little benefit from extending the crop portfolio beyond three crops. But not all crops are equally profitable every year. The black soil zone of Saskatchewan has the most diversification potential as moisture is a less limiting factor than elsewhere in the province. Even so, a two-crop portfolio of wheat-canola is stochastically dominant over other portfolios of commonly grown crops. Using the wheat-canola portfolio and the 1990/91 Saskatchewan crop insurance option of 60% of mean yields plus the high price option, crop insurance results in an approximately one in eleven payout, reducing but not eliminating the economic severity of major crop disasters; net cash available is only improved by $10,503.

References

Anderson JR, Dillon JL, Hardaker B (1977) Agricultural decision analysis. Iowa State Univ Press, Ames, Iowa

Newberry DMG, Stiglitz JE (1981) The theory of commodity price stabilization: a study in the economics of risk. Oxford Univ Press, 69-95

Saskatchewan Department of Agriculture and Food (1991) Cost of producing grain crops in Saskatchewan. Government of Saskatchewan, Regina, Canada

Schoney RA (1991) Top management farm business simulator and forward planning manual, version 6.15. Dept Agric Econ, Univ Saskatchewan, Saskatoon, Canada

Van Kooten GC, Schoney RA, Hayward KA (1986) An alternative approach to evaluation of goal hierarchies among farmers. West Agric Econ Assoc 11(2):40-49

Weisensel WP, Schoney RA (1989) An analysis of yield-price risk associated with specialty crops. West J Agric Econ 14(2):293-299

Wilson PN, Eidman VR (1983) An empirical test of the interval approach for estimating risk preferences. West J Agric Econ 8:170-182

Chapter 12

Crop Insurance and Agricultural Chemical Use

J.K. HOROWITZ and E. LICHTENBERG[1]

Deteriorating quality of the rural environment and agricultural resource base has become a growing source of concern in the United States. Federal farm programs are increasingly identified as having significant responsibility for the current state of affairs. The combination of target prices and set-asides characteristic of farm commodity programs is believed to give farmers significant incentives to increase their use of fertilizers and pesticides to increase yields (Reichelderfer 1990). A recent report by the National Academy of Sciences (National Research Council 1989) claims that base acreage requirements and other provisions of these programs severely limit flexibility in the use of crop rotations and other non-chemical means of managing soil fertility and pest problems. Favorable tax treatment of agriculture is thought to have accelerated irrigation development in the Midwest and therefore groundwater depletion and leaching of nitrates into groundwater supplies (Lichtenberg 1989 and Strange 1988). In recognition of these effects, the 1985 Farm Bill contained several provisions (the Conservation Reserve Program; conservation compliance; freezing base yields for deficiency payments) motivated by environmental concerns. The 1990 Farm Bill continued these and added several more (base acreage flexibility).

Another program with the potential to influence agricultural chemical use is federal crop insurance. Agriculture is an industry in which risk plays a substantial role in production decisions, including decisions about chemical use and cultivation practices which have potentially significant environmental effects. Thus crop insurance, which is aimed specifically at affecting risk, may also have an impact on environmental quality. The idea of using

---

[1]Department of Agricultural and Resource Economics, University of Maryland, College Park, MD 20742, USA.

subsidized crop insurance to promote reductions in pesticide use has been discussed at various intervals in policy circles. Like most linkages between government agricultural programs and environmental quality, however, the relationship between crop insurance and chemical use has received little, if any, empirical investigation. Carlson (1979) discusses conceptual problems in designing insurance against the risk of pest damage. Miranowski (1974) uses a farm-level simulation model to compare crop insurance with publicly provided scouting reports. The effects of insurance on farmers' actual chemical use decisions has not been studied to date.

In this paper, we study the relationships between crop insurance purchases and the use of nitrogen fertilizer and pesticides using farm-level data from corn farms in the U.S. Corn Belt, using the 1987 Farm Costs and Returns Survey. These two kinds of chemicals are of major environmental concern. Nitrogen runoff from farms accounts for increasingly large shares of nutrient pollution of major rivers, lakes and estuaries. Nitrate (largely from leaching of fertilizers and livestock wastes) has been found in over half of all rural drinking water wells (U.S. Environmental Protection Agency 1991). Pesticides have been associated with significant environmental problems on-farm, in surface runoff and groundwater and via aerial drift.

We focus on the influence of crop insurance choices on chemical use on land currently used for corn. A second possible effect of crop insurance subsidies would be to induce farmers to increase cultivation in riskier areas. If this cultivation required the use of heavy chemical applications, then increasing insurance subsidies could have the unintended result of increasing environmental damage from chemicals. The effects of crop insurance on cultivation of marginal lands and the accompanying chemical use are an interesting possibility, but will not be addressed in this study.

We begin with a simple theoretical model demonstrating the linkages between insurance purchases and agricultural chemical use. We then review the literature on agricultural input use under risk. Next, we discuss the data and estimation methods. Finally, we present some empirical results.

12.1 Crop Insurance and Input Demand

One might expect crop insurance to affect agricultural chemical use because it affords opportunities for moral hazard, in the sense that insured farmers could take fewer precautionary measures to reduce either the size or probability of crop damage (Arrow 1963 and Holmstrom 1979). While it is possible to structure contracts to minimize moral hazard as suggested by Quiggin in chapter 5 of this book, the Federal Crop Insurance Corporation (FCIC) appears not to have done so. Its principal aim has been to maximize participation, and it may have ignored moral hazard concerns in its attempts to meet this goal.

In what follows, we present a simple model to demonstrate the moral hazard effects of crop insurance. We distinguish these from selection effects, which may occur because farmers in riskier areas are more likely to participate in the crop insurance program. If these farmers also tend to use more or less of certain inputs than farmers in less risky areas, there will be a correlation between insurance and input use that is not induced by insurance coverage *per se*. We thus consider only changes in input use due to changes in insurance coverage, holding prices, weather and other sources of risk constant.

Consider a simple model of agricultural production with one input when crop insurance is available. Suppose that there are two states of nature, a bad one in which yields are sufficiently low that the farmer receives an insurance payment, and a good one in which yields are high and insurance does not pay off. Let the contract be such that in bad states of nature the farmer receives a payment based on an insured yield level y_B. For simplicity, assume that the price of the crop is constant across states of nature at p, so that the farmer receives a payment of py_B whenever yield falls below a critical level. Let the premium the farmer pays for insurance be proportional to the insured yield level y_B, and denote the price of insurance per unit of insured output by α. Suppose that output is a function of the amount of a single input applied, x. In the good state of nature, output is $y_G(x)$ and earnings from the sale of the crop are $py_G(x) - wx$, where w is the price of the input. Income in the good state of nature equals earnings from the sale of the crop minus the insurance premium, $py_G(x) - wx - \alpha y_B$, while income in the bad state of nature equals the crop insurance payment minus the insurance premium minus expenditure on the input, $(p - \alpha)y_B(x) - wx$.[2] Finally, suppose that the input alters the probability that the state of nature will be good, $\pi(x)$. We define the input as risk-increasing if $\pi'(x) < 0$ and as risk-reducing if $\pi'(x) > 0$ where the prime denotes differentiation.

The moral hazard effect is given by the marginal rate of substitution between insurance and input use. To see this, consider the choice of input use conditional on the choice of insurance. Given insurance coverage y_B, the farmer chooses x to

$$\max \pi(x)U(py_G(x) - wx - \alpha y_B + [1 - \pi(x)]U[(p - \alpha)y_B - wx] \quad (1)$$

The necessary condition for a maximum is:

$$\pi'[U_G - U_B] + \pi U'_G \cdot (py'_G - w) - (1 - \pi)U'_B \cdot (p - \alpha) = 0, \quad (2)$$

where the subscripts G and B denote the good and bad states of nature, respectively.

Equation (2) implicitly defines input use as a function of insurance coverage y_B. Differentiating equation (2) with respect to y_B gives the marginal impact of crop insurance on input use:

---

[2]The model presented here is a stylized version of the actual U.S. multiple-peril crop insurance program which includes a number of price and yield electives as described in chapter 8 of this book. We assume here the insured price is equal to the market price and that the "crop insurance payment" equals the indemnity payment plus the market price times actual yield if actual yield is less than the insured yield level.

$$\frac{dx}{dy_B} = \frac{\pi'[\alpha U'_G + (p - \alpha)U'_B]}{\Delta} + \frac{\alpha\pi U''_G \cdot (py'_G - w)}{\Delta} \tag{3}$$

$$+ \frac{w(1 - \pi)(p - \alpha)U''_B}{\Delta},$$

where $\Delta < 0$ is the second derivative of the objective function with respect to x.

If the farmer is risk neutral, the second two terms on the right hand side of equation (3) are zero, and the sign of dx/dy_B is determined solely by the sign of $\pi'(x)$. If the input is risk-increasing, $dx/dy_B > 0$; if it is risk-reducing, $dx/dy_B < 0$.

The second and third terms on the right-hand side of equation (3) are wealth effects, that is, they indicate the impacts of changes in insurance purchases on income in good and bad states of nature. In most cases, increasing x will increase income in good states because $py'_G - w > 0$; in bad states, however, increasing x will reduce income because expenditures increase by w while revenue (which equals the insurance payment) remains constant. Increased insurance coverage y_B reduces income in good states and thus increases the marginal utility of income, making additional income more attractive. The second term on the right-hand side of equation (3) will thus generally be positive. Additional insurance coverage increases income in bad states of nature, making additional losses of income less burdensome (by reducing the marginal utility of income) and thus increasing the attractiveness of spending on the input. The third term on the right-hand side of equation (3) will thus always be positive.

If the input is risk-increasing, equation (2) indicates that it will always be true that $py'_G - w > 0$. However, if the input is highly risk-reducing, it is possible that $py'_G - w < 0$. This will occur when the input reduces risk to such a great extent that its marginal effect on risk-reduction $\pi'[U_G - U_B]$ outweighs the marginal reduction in income $py_G' - w$. In this case, the decrease in the marginal utility of income caused by increased insurance coverage will lower the marginal value of risk-reduction, making the second term on the right-hand side of equation (3) negative.

When the input is risk-increasing, all three terms on the right-hand side of equation (3) are positive. When the input is risk-reducing, the first term is always negative, the third term is always positive and the second term may be positive or negative. Thus, increased insurance coverage will always lead to increased use of risk-increasing inputs, but may lead to increases or decreases in the use of risk-reducing inputs.

12.2 Risk Effects of Agricultural Chemicals

The preceding analysis demonstrates that the relationship between crop insurance and agricultural chemical use hinges on the risk effects of chemicals. In what follows, we review the literature on these risk effects.

Fertilizers are widely believed to be risk-increasing. Just and Pope (1979) looked at experimental data on corn yields and nitrogen fertilizer use and found nitrogen to be risk-increasing. The papers in Anderson and Hazell (1989) present evidence from a variety of developing countries showing that fertilizer increases variability in grain yields. Some contrary evidence also exists. Josephson and Zbeetnoff (1988) present evidence from yield studies indicating that fertilizer may be either risk-increasing or risk-reducing. SriRamaratnam, Bessler, Rister, Matocha and Novak (1987) elicited subjective probability distributions of yield from grain sorghum farmers in Texas. The majority believed that fertilizer was risk-reducing.

Pannell (1991) reviews the literature on pesticides and risk at length. Most economists have assumed that pesticides are risk-reducing. If pest infestation, damage per pest or pesticide effectiveness is the sole source of uncertainty, this will generally be true (see Feder 1979 for a formal treatment). Pannell notes that potential output and/or price are also frequently uncertain and that when they are the sole source of uncertainty, pesticides can be risk-increasing. With multiple sources of uncertainty, then, pesticides may be risk-reducing or risk-increasing (see Horowitz and Lichtenberg 1992 for a formal treatment).

12.3 Insurance and Agricultural Chemical Use

The relationship between chemical use and insurance coverage contains two possible effects: (1) selection effects that occur because the characteristics that lead farmers to select higher insurance coverage may also lead them to apply more of risk-reducing chemicals, and (2) moral hazard effects that occur because insurance reduces the risk farmers face, leading them to apply less of risk-reducing chemicals. Separating these effects is a challenging empirical problem but is important from a policy point of view because the effectiveness of crop insurance in reducing agricultural chemical use depends precisely on the moral hazard effect. On the other hand, the subsidy costs of any insurance program are determined primarily by adverse selection.

The farmer's decision to buy crop insurance is made prior to planting time and is conditioned on planting decisions, the distribution of prices and yields (conditional on input use), risk aversion, participation in commodity programs, and other opportunities to manage risk. The level of crop insurance coverage selected by the ith farm, y_i^*, is a function of these farm characteristics Z_i:

$$y_i^* = \alpha' Z_i + u_i. \tag{4}$$

The farmer purchases crop insurance if desired coverage y_i^* is positive and does not purchase insurance if desired coverage is negative. Let I_i be an indicator variable taking on a value of one if the farmer purchases insurance and zero if the farmer does not.

Chemical use decisions x_i are made throughout the growing season. They depend on these same farm characteristics Z_i and whether or not the farmer has purchased crop insurance coverage:

$$x_i = \beta' Z_i + \delta I_i + v_i. \tag{5}$$

In this model, α captures selection effects, while δ measures the moral hazard effect.

The sequential nature of the decision process implies that equations (5) and (6) might plausibly be treated as a recursive system in which the errors u_i and v_i are uncorrelated. We assume this to be true and thus estimate equation (6) for agricultural chemical use using a single-equation framework.

12.4 Data

Our estimates are based on farm-level data collected by the National Agricultural Statistical Service (NASS) in its Farm Costs and Returns Survey. The survey is an annual design survey of farmers in the United States who reported some corn acreage in 1987. We restricted our attention to a subsample of farmers in 10 states that constitute the heart of the Corn Belt. All counties in Indiana, Illinois and Iowa were included, plus selected counties in Kansas, Michigan, Minnesota, Missouri, Nebraska, South Dakota, and Wisconsin where dryland corn is a major crop. A total of 432 farms were included in this sample. Missing information limited us to 376 observations.

The data set contains expenditures, input use, farm debts and assets, and income in 1987. Summary statistics are provided in Table 1. The dependent variables that are relevant to our study are expenditures on crop and livestock insurance, applications of fertilizer and pesticides, and total expenditures on fertilizers and pesticides and the shares of these that were applied on corn. The measure of insurance purchases includes annual premiums for coverage of both crops and livestock, but not of motor vehicles or

Table 1. Data Used in the Analysis

	Mean	Standard Deviation	Minimum	Maximum
Insurance purchase (1 if yes)	0.47	0.50	0.00	1.00
Nitrogen per corn acre	113.73	0.29	0.00	360.40
Pesticide expenditures per acre	20.35	13.37	0.00	102.85
Total acres[a]	427.41	443.33	7.00	3230.00
Total acres in corn	184.18	199.23	3.0	1100.00
Percent of acres in soybeans[a]	0.29	0.22	0.00	0.96
Percent of acres in small grains[a]	0.10	0.11	0.00	0.70
Percent of acres in pasture[a]	0.15	0.22	0.00	0.92
Percent of corn acres irrigated	0.02	0.14	0.00	1.26
Percent of corn acres under low tillage[b]	0.27	0.36	0.00	1.00
Assets ($100,000)[c]	5.01	5.46	0.00	38.51
Value of livestock as % of assets[c]	0.07	0.11	0.00	0.87
Any acres rented for cash? (1 if yes)	0.46	0.50	0.00	1.00
% of acres operated rented for share	0.19	0.33	0.00	1.00
Any off-farm wages? (1 if yes)[d]	0.47	0.50	0.00	1.00
Any off-farm business? (1 if yes)[d]	0.17	0.37	0.00	1.00
Operator's age	50.91	12.98	25.00	87.00
Mean corn yield per acre[e]	87.92	22.31	34.83	118.46
Coefficient of variation, corn yield[e]	0.51	0.28	0.08	1.24
January-March precipitation	419.55	172.81	127.00	1132.00
April precipitation	205.61	90.00	16.0	429.00

[a]Total acres is all acres operated but not enrolled in a set-aside program.
[b]Includes cultivation classified as no till, strip till, and minimum till.
[c]Farmer's estimate of beginning of 1987 assets.
[d]Includes wages earned by operator and any other household member.
[e]County average.

buildings. Although most of this expenditure on insurance is likely to be FCIC insurance, other types of insurance (e.g., fire and hail) may be included. The fraction of the farm's assets in livestock is negatively related to crop insurance purchases, suggesting that insurance is purchased primarily for crops rather than for livestock.

The FCIC offers farmers a choice of coverage among several yield and price levels. Virtually all farmers choosing to purchase insurance from the FCIC choose the maximum yield and price offered. It is thus reasonable to treat the crop insurance decision as a dichotomous choice of whether or not to purchase insurance.

The quantity of nitrogen applied to corn is the product of application rate per acre, adjusted for the type of fertilizer applied, and the number of acres treated. We divided this by total corn acres to obtain an average per acre. This physical measure is likely to be more accurate than the expenditure data, since it includes manure applications, for which expenditure may be zero, and because the aggregate expenditure variable includes purchases of other elements such as phosphorus, potassium, lime, and micronutrients, which have quite different roles in the production process.

The physical measure of pesticide use was the number of corn acres treated with each of the major herbicides and insecticides, identified by chemical. Application rates were not available, and aggregation over chemicals was considered undesirable. In this case, expenditure data are likely to be more reliable, since they account implicitly for differences in effectiveness and application rates. Expenditures were converted to a per acre basis by dividing by corn acreage.

The set of explanatory variables in the input use equations for nitrogen and pesticide use was restricted to those that could reasonably be considered exogenous or predetermined. The shares of total acreage planted to alternative crops and used for pasture were included both as measures of the extent of diversification and because of agronomic effects (e.g., nitrogen fixing by soybeans). The fraction of total acreage under reduced tillage was included because pesticide use often differs according to tillage. Measures of production risk included the coefficient of variation of corn yield in the county in which the farm was located (see below for the source of these data), pre-plant (January through March) and at-plant (April) precipitation and the share of corn acreage that was irrigated.

Price data are not reported in this data set, except for state-level average hourly wage rates. Because prices are likely to vary little across the production region we study, little information is lost by focusing on expenditures and quantities.

Most of the financial variables are measured at the end of 1987, while the theory and econometric models are based on values at the beginning of the crop year. A few of the assets such as the value of livestock and crops on hand are recorded for the beginning of the year as well as the end. We assumed that changes in the value of buildings and machinery were similar across the region studied, allowing us to construct estimates of the value of assets as of January 1, 1987 before insurance purchase decisions are made. Debts were measured as of the end of 1987. Since this measure does not reflect the farm's debt position during the crop year, we did not use it. The percentage of assets in livestock was a measure of farm diversification.

The financial riskiness of an operation is also influenced by tenure arrangements. We included a dummy variable indicating whether any land was rented for cash, which should increase risk. We also included the percentage of acres operated that were rented for a share of the crop, a risk-reduction measure. Off-farm income diversification prospects were measured by a dummy variable indicating whether the farm earned off-farm wages during 1987 or operated an off-farm business. The age of the operator was included as a measure of human capital.

In addition to these data, we used county-level time series data on yields and insurance transactions, collected by FCIC. These were used to construct county level estimates of mean corn yield, which reflects average production potential, and the coefficient of variation of corn yield, which reflects production risk, as noted above.

The Farm Costs and Returns Survey is a design survey of farms rather than a random sample. Less commonly found types of farms are oversampled. To provide consistent estimates of the standard errors under this sampling design, we employ weighted least squares using the weights calculated by the National Agricultural Statistical Service for this purpose.

12.5 Results

Table 2 presents the parameter estimates for nitrogen applications and pesticide expenditures, both on a per-acre basis.

Crop insurance appears to have had a substantial influence on chemical use in the Corn Belt. Farmers who purchased crop insurance applied almost 20 pounds more nitrogen per acre and spent about $3.70 more per acre on pesticides. This corresponds to a 17 percent increase in nitrogen use over the average and an 18 percent increase in pesticide expenditures. Thus, subsidized crop insurance is likely to have detrimental effects on environmental quality, in that it is likely to promote even greater use of nitrogen fertilizer and pesticides.

By and large, the coefficients in the nitrogen equation are consistent with the conventional wisdom that nitrogen is risk-increasing. The risk of fertilizer burn is lower when pre-plant soil moisture is higher; the coefficient of January-March precipitation is correspondingly positive. Farmers with more assets, and thus presumably lower risk aversion, use more nitrogen. Farms operating a larger fraction of total acreage for share rent are presumably bearing less risk and are thus willing to use more of risk-increasing inputs. The only contrary piece of evidence is the fact that farmers operating some acreage for cash rent also tend to apply more nitrogen; they presumably bear greater risk, all other things being equal, and thus should tend to use less of risk-increasing inputs. However, this coefficient is significantly different from zero only at the 10 percent level.

Finally, nitrogen use per acre is greater in areas with higher average corn yield and thus greater potential productivity.

The coefficients in the pesticide equation convey mixed messages about the risk effects of pesticides in Corn Belt agriculture. On the one hand, pesticide use is lower in areas with

Table 2. Results of Chemical Use Regressions

Variable	Nitrogen per Acre	Pesticide Expenditure per Acre
Intercept	-75.77 (1.51)	39.888*** (3.52)
Insurance dummy	19.75*** (3.25)	3.69*** (2.68)
Total corn acreage	0.008 (0.335)	-0.006 (1.03)
County mean corn yield per acre	0.145*** (4.43)	-0.005 (0.69)
Coefficient of variation of corn yield per acre	35.29 (1.27)	-19.10*** (3.03)
Share of farm acreage in soybeans	28.87 (0.64)	-8.50 (0.83)
Squared share of acreage in soybeans	-9.09 (0.13)	19.44 (1.26)
Share of farm acreage in small grains	10.86 (0.68)	-0.08 (0.02)
Share of farm acreage in pasture	-9.05 (0.42)	9.86** (2.04)
Share of farm acreage under reduced tillage	-2.56 (0.27)	0.90 (0.42)
Share of corn acreage irrigated	24.01 (0.64)	-5.60 (0.66)
January-March precipitation	0.040** (2.06)	-0.01*** (2.92)
April precipitation	-0.03 (0.64)	-0.02* (1.76)
Assets ($100,000)	2.99*** (2.93)	0.49** (2.13)
Value of livestock as share of assets	-9.49 (0.35)	3.27 (0.53)
Share of acreage operated on shares	22.10** (2.07)	-1.86 (0.77)
Rented some acreage for cash (1 if yes)	13.57* (1.76)	2.59 (1.48)
Age of operator	-0.23 (0.87)	-0.03 (0.47)
Received off-farm wages (1 if yes)	6.85 (1.14)	-1.67 (1.23)
R^2	0.30	0.16

Asympotic t-ratios in parentheses. Sample size = 375.

***Significantly different from zero at the 1 percent level.
**Significantly different from zero at the 5 percent level.
*Significantly different from zero at the 10 percent level.

greater yield variability. As Pannell (1991) has noted, pesticides tend to be risk-increasing in the presence of significant output variability. On the other hand, pesticide use is lower when pre-plant and at-plant soil moisture are greater. Since yield uncertainty tends to be less when the latter are greater, these coefficients are more consistent with pesticides as risk-reducing inputs. In general, however, the risk effects of inputs tend to be complex. For example, Horowitz and Lichtenberg (1992) have shown that when there are multiple sources of uncertainty, it is difficult to characterize inputs as either strictly risk-increasing or risk-reducing, and inputs may be risk-increasing for some levels of output or under some realizations of random conditions and risk-reducing under others.

Finally, farmers with more assets spend more on pesticides, as one would expect.

12.6 Final Remarks

Negative externalities from agricultural production are increasingly a source of concern. Agriculture has come to account for rising shares of nutrient pollution in waterways. Nitrate from fertilizers and animal wastes and pesticide residues have emerged as major sources of groundwater pollution (U.S. Environmental Protection Agency 1991). Pesticide runoff and aerial drift have frequently caused significant environmental problems.

In recent years, the effects of agricultural programs on input use have been the subject of considerable discussion. The proposition that farm commodity programs promote more intensive chemical use has gained wide currency; however, empirical evidence has been scant (Reichelderfer 1990). Crop insurance has also been thought to influence farm chemical use, especially pesticides, and crop insurance subsidies have been discussed as a means of effecting reductions in pesticide application. This paper has investigated the empirical relationship between agricultural chemical use and crop insurance. Our results imply that crop insurance subsidies will result in greater use of both nitrogen fertilizer and pesticides and will thus have negative impacts on environmental quality. Farmers purchasing crop insurance apply almost 20 percent more nitrogen fertilizer per acre and spend nearly 20 percent more per acre on pesticides. Our results thus confirm Carlson's (1979) and Miranowski's (1974) skepticism about crop insurance as a tool for environmental policy.

References

Anderson JR, Hazell PRB (1989) Variability in grain yields: implications for agricultural research. Johns Hopkins Univ Press, Baltimore

Arrow KJ (1963) Uncertainty and the welfare economics of medical care. Amer Econ Rev 53:941-974

Carlson GA (1979) Insurance, information and organizational options in pest management. Annu Rev Phytopathology 17:149-161

Feder G (1979) Pesticides, information, and pest management under uncertainty. Amer J Agr Econ 61:97-103

Holmstrom B (1979) Moral hazard and observability. Bell J Econ 10:74-91

Horowitz JK, Lichtenberg E (1992) Are pesticides risk-reducing? Dept Agr Resource Econ, Univ Maryland

Josephson RM, Zbeetnoff D (1988) The value of probability distribution information for fertilizer application decisions. Canad J Agr Econ 36:837-844

Just R, Pope R (1979) Production function estimation and related risk considerations. Amer J Agr Econ 61:276-284

Lichtenberg E (1989) Land quality, irrigation development, and cropping patterns in the northern high plains. Amer J Agr Econ 71:187-194

Miranowski JA (1974) Crop insurance and information services to control use of pesticides. Office of Research and

Development, U.S. Environmental Protection Agency, Washington DC

National Research Council (1989) Alternative agriculture. National Academy Press, Washington DC

Pannell DJ (1991) Pests and pesticides, risk and risk aversion. Agr Econ 5:361-383

Reichelderfer K (1990) Environmental protection and agricultural support: are tradeoffs necessary? In: Allen K (ed.) Agricultural policies in a new decade. Resources for the future, Washington DC, 201-230

SriRamaratnam S, Bessler D, Rister ME, Matocha J, Novak J (1987) Fertilization under uncertainty: an analysis based on producer yield expectations. Amer J Agr Econ 69:349-357

Strange M (1988) Family farming. University of Nebraska Press, Lincoln

U.S. Environmental Protection Agency (1991) Another look: national survey of pesticides in drinking water wells phase II report. EPA 579/9=91-021, Washington DC

Chapter 13

Crop Insurance: Its Influence on Land and Input Use Decisions in Saskatchewan

W.P. WEISENSEL, W.H. FURTAN[1] and A. SCHMITZ[2]

13.1 Introduction

Saskatchewan has over 40 percent of all the arable farm land in Canada. The major crop grown in the province is wheat with around 20 million acres planted annually. Other major annual crops are barley, canola and flax. Numerous other smaller acreage crops like oats, peas, lentils and mustard are also grown. Each of these crops are insurable, and farmers must make the decision to insure in the months of March and April.

In Saskatchewan, crop insurance is an important stabilization policy. Its importance is indicated by the fact that more than 75 percent of producers currently participate. This participation rate is high, compared to rates in other countries. However, the high level of participation is easily explained by the fact that the provincial and federal governments provide a 50 percent subsidy for crop insurance premiums and a 100 percent subsidy for program administration. Consequently, participation in crop insurance results in a net transfer to grain production on the prairies for the majority of farm operations.

In recent years, there has been a growing concern that current land use and agronomic practices have been detrimental to the preservation of Saskatchewan soils (Rennie 1986). Crop insurance, which is perceived as a program that influences agronomic decisions, is at the center of this concern (Agriculture

[1]Department of Agricultural and Resource Economics, University of Saskatchewan, Saskatoon, Saskatchewan, Canada.

[2]Department of Agricultural and Resource Economics, University of California, Berkeley, California, USA.

Canada 1989). However, little is known about how agricultural practices on the prairies have been affected by the Canada-Saskatchewan All-Risk Crop Insurance program. The purpose of this paper will be to examine the possible effects that crop insurance has had on land and input use. This paper first presents a conceptual model which explains how crop insurance influences the farm management decisions of a participator. In the second section, the data used to test the implications of the conceptual model are described. In the third section, the results of the empirical analysis are discussed.

Two methods of analysis are applied in this paper. Since farmers have different wealth levels, their degrees of risk aversion are likely to differ. This, in turn, may cause differences in the behavior of firms regarding their land and input use decisions. First, we use a method of analysis based on the perception of how crop insurance has changed farmers' decisions. This is done by asking farmers particular questions regarding their current behavior and comparing it to how they would have behaved in the absence of crop insurance. The second method of analysis compares the agronomic decisions of participators and non-participators and uses this information to predict the decision to participate in crop insurance.

13.2 Conceptual Model

A partial equilibrium model is used to explain the influences that crop insurance has on the land use decisions of two representative producers. We assume that the two producers are identical in all aspects except their levels of risk aversion. In Fig. 1, we illustrate the planning supply function for both producers prior to their participation in crop insurance, given that they have identical cost structures. The risk neutral producer plans on the supply curve S while the risk averse producer makes his decisions on the supply curve S^A. Therefore, the vertical difference between the two supply

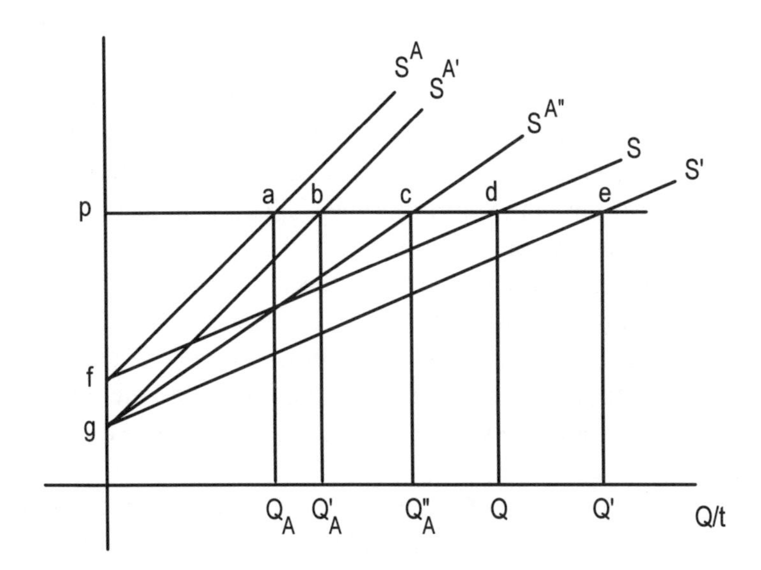

Fig. 1. Modelling farm management responses to crop insurance

curves represents the risk premium of the risk averse producer.[3] Given an expected price of P, the risk averse individual plans to produce output Q_A while the risk neutral individual plans to produce Q. Thus, the risk neutral producer uses more inputs than the risk averse farmer. Suppose now that a subsidized crop insurance program is introduced and both producers participate. How does this influence the decisions of each of these producers?

A risk neutral producer is influenced by changes in the expected value of an activity, regardless of the variance associated

[3]This risk premium is a function of the producer's level of risk aversion as well as the variance of expected revenue (Newberry and Stiglitz 1981).

with each of the activities. Therefore, in the risk neutral case, it is important to determine how a subsidized crop insurance program could influence rational decision making. Recent research examining the Saskatchewan Crop Insurance program has shown that the expected subsidy associated with insuring stubble crops is greater than that for summerfallow crops for all regions where area yields are lower for stubble crops than summerfallow crops (Weisensel, Furtan and Schmitz 1990).[4] This bias occurs because the Saskatchewan Crop Insurance Corporation charges the same premium rate for a given crop regardless of whether it is sown on stubble or summerfallow.[5] Therefore, because the guaranteed yield is a percentage of the area average yield, a lower yielding stubble crop is more likely to receive an indemnity than the same crop grown on summerfallow, even though it is charged the same premium rate.

In Fig. 1, the net effect of this bias is shown as a horizontal shift of the risk neutral individual's supply curve from S to S′. This results in an increase in output from Q to Q′ due to greater use of stubble cropping.[6] Similarly, the risk averse producer is also affected by changes in relative profitability. Therefore, the bias of the crop insurance program toward stubble cropping will also result in a horizontal shift in supply from S^A to $S^{A'}$. However, in addition

---

[4]Area average yields for summerfallow crops are greater than stubble crops in most areas of the province since summerfallow crops are grown on two seasons of precipitation while stubble crops are grown on a single year of precipitation. The discrepancy is greatest in the brown soil zone, where seasonal precipitation is most limiting, followed by the dark brown soil zone. Some regions of the black soil zone have equivalent summerfallow and stubble yields.

[5]Crops are insured at either 60 or 70 percent of the area average yield. The premiums are not sensitive to the differences in the coefficient of variation of summerfallow and stubble yields.

[6]The size and direction of this shift is an empirical question which will depend upon the relative profitability of extended cropping. For instance, if crop insurance was biased toward summerfallow crops rather than stubble crops, it is conceivable that the supply curve could shift in as a response to crop insurance.

to the changes in the expected value of insurance, the crop insurance program also reduces the revenue risk associated with production. Therefore, the size of the risk premium of the risk averse producer is reduced, resulting in a radial shift in his supply curve from $S^{A'}$ to $S^{A''}$. This shift results in an increase of output equal to the difference between Q_A' and Q_A''. Theoretically, this second increase in output can be a result of two separate factors: (1) an increase in variable input use, and (2) a further increase in the intensity of stubble cropping.[7] However, note that the increases in variable input use and/or stubble cropping do not result in an expected supply equivalent to that of the risk neutral producer. They simply reduce the differences in output between the two producers.[8] This has implications for the empirical work to follow.

It is not difficult to expand the model to incorporate more crops. In this case, risk neutral producers will choose those crops which yield them the highest expected return regardless of the risk associated with them. Crop insurance will not influence the crop choice for risk neutral producers provided the insurance coverage offered on each of the crops does not create a bias towards any particular crop. For instance, if crop insurance is more biased towards wheat than barley, the supply curve for wheat will shift out while the supply curve for barley will shift in. The size of these shifts will depend upon the size of the bias and the substitutability

---

[7]We avoid the debate of whether an input is risk increasing or risk reducing (Nelson and Loehman 1987, Just and Pope 1978). We assume that inputs in prairie agriculture are risk increasing since it is unlikely that there are any significant inputs in prairie agriculture which are risk reducing.

[8]Even a pareto optimal insurance contract (Nelson and Loehman 1987) does not eliminate price risk. Therefore, the risk premium of the risk averse producer must still be positive.

of wheat for barley. Whether this case actually exists for any crops is an empirical question.[9]

In contrast, risk averse producers will trade off higher expected profits to reduce their risk. Consequently, their crop selections are more likely to be affected by a crop insurance program, particularly if some of the crop choices available to the producer are not covered by crop insurance. This is an empirical question which depends on the expected profitability and risk of the crops in question, as well as the level of risk aversion of the producer (Weisensel and Schoney 1989).

Finally, it is important to note that crop insurance may actually encourage more crops because it allows the producer to insure tracts of land with different crops separately. This incentive is created by the fact that crop insurance treats each crop insured on a given farm separately. For example, a producer farming four quarter sections could insure each quarter section as a separate unit by seeding a different crop on each quarter section.

The discussion above focuses on how crop insurance may influence the decisions of a participator on a tract of land already used in the production of grain. However, the influence that crop insurance may have on the cultivation of marginal land is also important. Fig. 1 also provides some insight into this question. Conceptually, each of the responses to crop insurance outlined in the paragraphs above result in an increase in quasi-rent in the production of grain. For the risk neutral producer, the increase in quasi-rent is the area fdeg. In comparison, the change in quasi-rent for the risk averse producer is the area facg. These areas represent the increased returns to the fixed factors of production, one of

---

[9]In addition, from a farm income perspective, crop selection will also depend upon the risk diversifying effects of adding more crops to the crop portfolio (see Anderson, Dillon and Hardaker 1977). This may have important implications for specialty crops which are not highly correlated with crops traditionally grown in Saskatchewan (Weisensel and Schoney 1989).

which is rent to land. Therefore, the increases in quasi-rent favor land used in grain production as opposed to other alternative uses.[10]

13.3 Data

There are significant regional land quality differences in Saskatchewan (Schoney, Thorson and Weisensel 1988).[11] Therefore, it seems likely that crop insurance will affect farmers in separate regions differently. To incorporate these possible differences, the farm survey was conducted in four regions: Chaplin-Central Butte (south, low quality); Moose Jaw (south, high quality); Melfort (north, high quality); and Meadow Lake (north, low quality). In the survey, information was collected from producers about their basic farming practices. The survey included questions about input use, possible land clearing and other agronomic practices. Its purpose was to provide data to test whether there are differences in the way participators and non-participators farm their land.[12]

---

[10]Even if there is no supply response (shift) due to a bias in crop insurance, a subsidized program will still result in a downward shift in the average total cost curve and an increase in land rent in the production of grain. For example, if no biases existed in a subsidized crop insurance program (for instance stubble versus summerfallow), the supply curve for the risk neutral producer would not shift as shown in Fig. 1. However, the rent to land in grain production would still increase because the average total cost curve shifts down.

[11]Traditionally, the agricultural region of Saskatchewan has been divided into three climatic zones known as the brown, dark brown, and black soil zones. The brown soil zone corresponds to the southwestern corner of the province where rainfall is most limiting. The black soil zone is in the northeastern region of the grain belt and receives the greatest amount of precipitation. The dark brown soil zone falls between the black and brown soil zones.

[12]A total of 250 interviews were conducted from July 17th to August 18th, 1989. In each of the areas, a number of farmers refused to be interviewed. The rate of refusal was highly correlated with the intensity of livestock in a given area. The

13.4 Sample Characteristics by Region

In this section some of the basic farm data are described. Tables 1 to 4 provide the summary statistics for each region for: the farmer's age; farm size; the percentage of gross income contributed by livestock operations; the level of high school education; farm solvency (debt/asset ratio); and the proportion of sample farmers who participate in crop insurance.

When one compares the results in Tables 1 to 4, a number of points become clear. The average age of farmer is stable across the four regions at around 46 years. Average farm size is greater in the two southern regions than in the two northern regions, and farm size is greater in the regions of lower soil quality than in the high quality land regions for both the northern and southern areas. In fact, when one compares Chaplin-Central Butte (Table 1) to Moose Jaw (Table 2), one can see that, on the basis of cultivated acres, farms in the Chaplin-Central Butte region tend to be larger. One would expect this since it takes a larger farm to support a family unit on land of lower quality.

Livestock tends to be a much more important component of the agriculture sector in the regions with lower and less homogeneous land quality (Chaplin-Central Butte and Meadow Lake). Thus Tables 1 and 2 reflect the fact that farms in each of these areas tend to have a substantial number of acres in permanent pasture or hay. This study examines livestock because of the link between livestock and marginal lands. When farmers improve their marginal land, they reduce their livestock numbers and use this land for other purposes.

---

timing of the survey was somewhat inconvenient since haying operations were in full swing. Only 10 farmers refused outright because they were disgusted with the crop insurance program and felt the survey process was useless. Refusals of this type occurred most frequently in the southern areas where 7 farmers refused to be interviewed.

Table 1. Sample characteristics of the Chaplin-Central Butte region

Statistic	Age	Total Acres	Cultivated Acres	Pasture	Gross Proportion Livestock Receipts	H.S. Educ.	Debt/Asset Ratio
Mean	45.39	2179	1543	557	0.27	10.73	10.25
Std. Dev.	13.42	1726	1130	926	0.25	1.59	0.25
Minimum	23	320	280	0	0	8	0
Maximum	75	9800	6300	5000	0.9	12	0.825

	Participators	Non-Participators	Producers with Livestock
Proportion	63.8%	36.2%	63.8%

Table 2. Sample characteristics of the Moose Jaw region

Statistic	Age	Total Acres	Cultivated Acres	Pasture	Gross Proportion Livestock Receipts	H.S. Educ.	Debt/Asset Ratio
Mean	47.09	1437	1320	64	0.08	11.09	0.15
Std. Dev.	12.67	1013	983	151	0.18	1.26	0.18
Minimum	26	0	0	0	0	7	0
Maximum	74	5940	5800	660	0.65	12	0.75

	Participators	Non-Participators	Producers with Livestock
Proportion	73.8%	26.2%	21.5%

Table 3. Sample characteristics of the Meadow Lake region

Statistic	Age	Total Acres	Cultivated Acres	Pasture	Gross Proportion Livestock Receipts	H.S. Educ.	Debt/Asset Ratio
Mean	47.97	1215	943	254	0.30	9.29	0.30
Std. Dev.	13.24	1040	858	413	0.33	2.34	0.36
Minimum	25	0	0	0	0	3	0
Maximum	75	5000	5000	1950	1	12	1.5

	Participators	Non-Participators	Producers with Livestock
Proportion	38.3%	61.3%	60.0%

Table 4. Sample characteristics of the Melfort region

Statistic	Age	Total Acres	Cultivated Acres	Pasture	Gross Proportion Livestock Receipts	H.S. Educ.	Debt/Asset Ratio
Mean	46.87	1135	1011	42	0.07	11.05	0.35
Std. Dev.	12.99	686	655	88	0.14	1.58	0.44
Minimum	24	0	0	0	0	7	0
Maximum	82	4160	3800	580	1	12	2

	Participators	Non-Participators	Producers with Livestock
Proportion	41.7%	56.7%	26.7%

13.5 Empirical Analysis

The conceptual model provides a number of hypotheses that are testable using empirical data. First, it suggests that risk averse participators are likely to use more inputs after participation in crop insurance than before. Second, it suggests that participators are more likely to stubble crop after joining the program. Third, it suggests that participators are likely to seed more crops, or to select the fields that they allocate to specific crops carefully to enhance the level of insurance for the farm operation. Finally, since crop insurance premiums are subsidized dollar for dollar by the government, crop insurance provides a subsidy to land used to produce grain.[13] Therefore, crop insurance may encourage the cultivation of marginal land in Saskatchewan. This incentive may reduce the attraction of livestock production (for example, beef) as an alternative enterprise on many farms. Below we use empirical data to test each of these hypotheses using a number of different statistical approaches.

Initially, we discuss the results of the survey which elicits information from farmers about how they feel that crop insurance influences their farm management decisions. Secondly, we examine whether the introduction of crop insurance has had a measurable impact on the cultivation of marginal land and the amount of summerfallow in the province. These two approaches address the question of whether a producer operates a farm differently, in terms of the quantity of inputs used and the amount of land stubble cropped, before and after enrolling in the crop insurance program. Therefore, this analysis provides a direct test of the hypotheses of the conceptual model, since it deals with how crop insurance affects the decisions of a particular producer and how these decisions affect aggregate provincial data.

The second question dealt with in this section is: do farmers who participate in crop insurance operate their farms

[13]The Saskatchewan Crop Insurance Corporation is currently adding forage and pasture insurance in some areas.

differently from those who do not participate? This comparison does not necessarily identify those changes in farming practices which may be due to crop insurance, but it does provide a benchmark comparison of two separate groups of producers: those who have no crop insurance, and those who have already adjusted to the crop insurance program. There are many reasons why the two groups of farmers may operate differently. For example, they may have different risk preferences, or they may farm on different soil types.[14] As a result, one must be cautious regarding the conclusions one draws concerning the impact of crop insurance on farming and land use patterns in comparing these two groups. To do this more accurately, one needs individual farm data before and after program participation. However, because this type of data is unavailable, our analysis relies, in part, on comparisons of the farming practices of participators and non-participators, ex-poste. In other words, the participators' adjustment to crop insurance has already occurred.

13.6 Analysis of Data Dealing with Farmers' Perceptions of Crop Insurance

In this section, a scaling procedure is used to elicit a participator's behaviorial response to the crop insurance program. One of the most widely used scaling procedures is called Likert scaling (Walizer and Wiener 1978). With Likert scaling, questions are designed to assess attitudes, beliefs or behaviors. For each of the surveyed regions, Table 5 presents the results of the Likert scaling questions about the number of crops seeded (CROPS), the level of variable input use (VARINPUTS), and the fields allocated to specific crops (FIELDS), in columns one through four, respectively.

---

[14]Recent research at the University of Saskatchewan suggests that crop insurance participators are more risk averse than non-participators (Weisensel and Schoney 1990).

Table 5. Influences of crop insurance on farm management decisions

	STUBBLE	CROPS[a]	VARINPUTS[a]	FIELDS[b]
CHAPLIN-CENTRAL BUTTE				
Mean	NA	3.13	3.09	1.44
Standard Deviation	NA	0.34	0.29	0.84
Proportion	10.0%	12.0%	10.0%	24.0%
MOOSE JAW				
Mean	NA	3.16	3.05	1.77
Standard Deviation	NA	0.41	0.22	1.17
Proportion	14.7%	14.7%	4.9%	34.4%
MEADOW LAKE				
Mean	NA	3.00	3.03	1.14
Standard Deviation	NA	0.00	0.16	1.42
Proportion	0.0%	0.0%	2.8%	13.8%
MELFORT				
Mean	NA	3.06	3.06	1.16
Standard Deviation	NA	0.31	0.31	0.50
Proportion	2.0%	4.0%	4.0%	10.0%

Source: Farm Survey

[a]The results are based on responses to a Likert score question where 1 through 5 represent markedly decrease, moderately decrease, no influence, moderately increase, and markedly increase, respectively. The proportion row illustrates the percentage of producers who gave a response greater than 3.

[b]The results are based on responses to a Likert score question where 1 through 4 represent no influence, small influence, moderate influence, and major influence, respectively. The proportion row illustrates the percentage of producers who gave a response other than 1.

One of the major concerns regarding crop insurance in Saskatchewan is how crop insurance affects the decision to stubble crop.[15] The first column of Table 5 (STUBBLE) indicates that crop insurance does influence some farmers' decisions to stubble crop.[16] However, the results suggest that crop insurance has a greater influence on the stubble cropping decision in the two southern regions. In the Chaplin-Central Butte and Moose Jaw regions, 10.0 and 14.7 percent of the participators stated that their participation in crop insurance increases the amount of land they stubble crop. Only one producer admitted this in the Melfort region while no producers perceived a difference in the Meadow Lake region. The differences in responses noted between the northern and southern regions is probably explained by the fact that stubble cropping is a more common and accepted practice in the north than in the south.[17] Since the risk of stubble cropping in the south is substantially greater than in the north, crop insurance has more influence where the risk is greater.

The third and fourth columns of Table 5 refer to how crop insurance has influenced a participator's decisions regarding the number of crops seeded (CROPS) and the level of variable input use (VARINPUTS). Each of the above decisions were rated by the participator using a Likert scale where a response of 1 through 5

---

[15]Soil Conservation Canada maintains that crop insurance reduces the incentive to stubble crop. In contrast, some maintain crop insurance increases the likelihood of stubble cropping and therefore distorts grain production in Saskatchewan.

[16]The proportion row for each region represents the percentage of farmers in the region who feel that crop insurance does increase the amount of land they stubble crop, regardless of the size of the influence.

[17]Technological advances in the use of fertilizers and chemicals for stubble cropping have been more favorable in northern cropping areas where moisture is less limiting. Since yield guarantees based on 15 year average stubble yields do not reflect current technology, stubble cropping decisions are not as likely to be affected where the technological change has been the greatest.

referred to markedly decrease, moderately decrease, no influence, moderately increase, and markedly increase, respectively.

Currently, the Saskatchewan Crop Insurance Program requires producers to insure the total production of a particular crop across the whole farm unit. Consequently, it is argued that many producers split their farm operations by seeding more crops to enhance their level of coverage. The results in Table 5 support these speculations. It is interesting to note that more than 12.0 percent of the participators in the two southern regions admit that crop insurance causes them to moderately or markedly increase the number of crops they seed. This compares to less than 4 percent of the participators in the two northern regions.[18]

The conceptual model of crop insurance indicates that risk averse producers may use more inputs with crop insurance than without. The results in Table 5 provide some support for the conclusions of the conceptual model. In each of the four regions, the mean value of the response is slightly greater than 3.0. Again, it is interesting to note that more participators in the south responded positively to this question than participators in the north. Theoretically, one could explain this result if crop insurance reduces income variability less in the north than in the south. This would result in a greater radial shift in the supply curve of risk averse producers in southern regions when compared to northern regions (see Fig. 1). Our conclusion, based on discussions with producers in northern areas, is that crop insurance is regarded as a worst case scenario since indemnities do not cover the variable costs of production.[19] In contrast, crop insurance coverage in

---

[18]This is a rather interesting result when one considers that northern producers, who receive greater annual precipitation, have more crop options than southern farmers. The proportion row indicates the proportion of producers who feel crop insurance has had a positive impact upon CROPS, VARINPUTS, INVEST and FIELDS.

[19]This is particularly true when market prices are low. This seems to support the contention that crop insurance does not reduce income instability in the north to the same degree as it does in the south. In addition, empirical evidence to support

southern areas covers the variable costs of crop production in most cases.

Another method of increasing the probability of collecting indemnities with crop insurance is to select carefully the fields a producer allocates to the seeding of specific crops. One example of this is a farmer who has two or more tracts of land which are located in geographic or climatically different regions.[20] In order to maximize his coverage this producer will seed different crops in each region so low yields in one region will not be offset by high yields in another region. The FIELDS column in Table 5 presents the results of a question to participators concerning whether or not they feel crop insurance has an influence on the fields they allocate to specific crops. The Likert scale used to measure the importance of this factor ranks from 1 to 4, representing no influence, small influence, moderate influence, and major influence, respectively. Overall, the field allocation question appears to be the most important factor of all those shown in Table 5. Field allocation appears to be more important in the southern regions than in the northern.[21] In the southern areas, roughly 30 percent of participators indicated that crop insurance has at least a small influence on the fields they allocate to specific crops. This number drops to 12 percent for the two northern areas. One possible explanation for this discrepancy may be that northern farms tend to have smaller more concentrated land bases than southern farmers (Tables 1 through 4). Therefore, selecting acreage for specific crops in the manner described above may have less benefit to a northern producer.

---

this is provided in Weisensel, Furtan and Schmitz 1990.

[20]One weather pattern (drought) is unlikely to affect the separate regions in the same way. It is more likely, for example, that one of the areas will be a total loss while the other yields an average crop.

[21]The proportion row for each region refers to the number of participators who indicated that crop insurance has even a small influence on the fields they allocate to specific crops.

The above discussion focuses on farmers' perceptions about the influence that crop insurance has on farm management decisions. The results suggest that some producers do feel that crop insurance has influenced the way they operate their farm. However, this does not necessarily mean that crop insurance has a major influence at the aggregate level, since the proportion of producers who felt they were influenced combined with the marginal impacts of these influences may be negligible in the aggregate. Therefore, in the next few paragraphs, possible influences that crop insurance has on two major land use practices are examined using aggregate data. These are the proportion of summerfallow acreage and the ratio of improved to unimproved land in the province of Saskatchewan.

In Table 6, the seeded acreage of the major crops and summerfallow is given for Saskatchewan. The data are divided into two periods, 1955-64 (before crop insurance) and 1975-84 (after crop insurance).[22] The ratio of total seeded acres to summerfallow acres is calculated. In the period before crop insurance, the ratio was approximately 1.45 with the exception of 1961 which was a particularly dry year in Saskatchewan. In the after crop insurance period, the ratio of seeded acres to summerfallow acres increases after 1977. Therefore, it appears from this data that the increase in the ratio of seeded acres to summerfallow acres was independent of the introduction of the crop insurance program.

The data in Table 7 tell a similar story. This table shows the ratio of improved farm land (that is, crop land, summerfallow and improved pasture) to unimproved farm land. Over the thirty-five year period between 1951-1986, this ratio steadily increased but did not rise sharply after the introduction of crop insurance. This is not surprising since the provincial government has maintained a number of farm programs over this period that subsidized the cost of clearing, breaking and draining of farm

---

[22]Crop insurance was introduced in 1961. However, it was 1974 before a large number of farmers joined.

Table 6. Ratio of total seeded acreage and summerfallow before and after crop insurance: 1955-64 and 1975-84

Year	All	Oats	Barley	Rye	Flax	Canola	Tame Hay	Summer Fallow	No. of farmers participating	Ratio of seeded to Fallow Acres
1955	14100	3654	3846	457	1030	123	621	14284	0	1.668
1956	14569	3042	3027	300	1710	297	647	14194	0	1.662
1957	13800	2033	3791	268	2025	520	820	14900	0	1.560
1958	14200	2080	3800	247	1420	535	984	15800	0	1.472
1959	15800	1876	2960	253	871	165	944	16000	0	1.429
1960	15800	1876	2960	253	871	165	944	16000	0	1.478
1961	16082	1492	1839	239	941	374	1052	17180	194	1.281
1962	17388	2712	1629	292	389	167	1052	17180	194	1.383
1963	17920	2216	1930	336	506	210	1110	16898	2235	1.433
1964	19200	1469	1400	320	521	303	1165	16815	2357	1.449
1975	15200	1850	3500	325	450	2000	2000	17900	31411	1.414
1976	17700	1600	2950	267	200	750	2200	17800	38209	1.441
1977	16200	1550	3800	250	600	1450	2150	17800	39143	1.460
1978	17100	1200	3100	350	450	2800	2100	17000	47156	1.594
1979	17100	1000	2600	360	800	3300	2000	17100	43032	1.588
1980	17600	900	3250	275	400	2000	1900	17600	40154	1.459
1981	19800	1100	3700	500	350	1350	1700	16500	42850	1.727
1982	19600	1100	3450	495	500	1500	1750	16300	46259	1.742
1983	20700	850	2750	470	250	2100	1750	15900	44469	1.815
1984	20000	750	3200	390	650	3200	1800	14800	43000	2.026

Source: Saskatchewan Agricultural Statistics Handbook, 1987 and Saskatchewan Crop Insurance Corporation.

Table 7. Ratio of improved to unimproved farm land in Saskatchewan: 1951-1986

Year	Farm Land	Improved Land	Unimproved Land	Improved/Unimproved
1951	61,663,195	38,806,770	22,856,425	1.697
1956	62,793,979	40,506,000	22,287,979	1.817
1961	64,415,518	43,117,813	21,297,705	2.024
1966	65,409,363	45,468,776	19,940,587	2.280
1971	65,056,875	46,426,487	18,630,388	2.491
1976	65,511,431	46,774,551	18,736,920	2.496
1981	65,564,000	48,639,871	16,924,129	2.873
1986	65,728,443	49,530,723	16,197,720	3.057

Source: Saskatchewan Agricultural Statistics Handbook, 1987

lands. Therefore, while crop insurance did not hinder the development of unimproved land, it appears that the process was well underway before crop insurance became an important farm policy. This statement is supported by recent research by Van Kooten and Schmitz (1989) which indicated that crop insurance, by itself, was not a major factor in the cultivation of marginal land.[23]

13.7 Practices of Participators and Non-Participators

An empirical model was constructed to test for the effect of differences in farm practices on the probability of participation. This model predicts whether a farmer will be a participator or not depending upon specific farm practices. Therefore, the model

[23]Recent work by Rosaasen, Eley and Lokken (1990) indicates that the direction of most government programs, including crop insurance, is towards the cultivation of marginal land. However, they imply that other government programs, like the Western Grain Transportation Act and the Western Grain Stabilization Act, were much more important than crop insurance in the magnitude of the influence.

empirical section: do the two groups of farmers (participators and non-participators) operate their farms differently with respect to some measured land and input use variables?

The probit model estimated appears as follows:

$$PR = f(ST/AC, ACBR, VAIP, SNTP, CHFL, STCP, LEGM, CPNO, LDUM)$$

where the dependent variable PR takes a value of 1 if a producer is a participator and 0 otherwise. The variable ST/AC is the ratio of stubbled acres to total seeded acres, ACBR is the number of improved acres (that is, drained, broken or cleared) since 1972, VAIP is the $/acre spent on chemical and fertilizer, SNTP is a dummy variable which is equal to 1 if the farmer stubble snowtraps, CHFL is a dummy variable which is equal to 1 if the farmer uses chem-fallow, STCP is a dummy variable which is equal to 1 if the farmer practices strip cropping, LEGM is a dummy variable which is equal to 1 if the farmer has a legume in the rotation, CPNO is the number of crops grown on the farm, and LDUM is a dummy variable which is equal to 1 if the farm has livestock income. The parameter estimates on each of the variables are provided in Table 8. When one compares the stubble cropping practices of participators and non-participators, the results of the probit do not provide any conclusive evidence. In two of the regions, the results seem to suggest that participators tend to have less of their total seeded acres in stubble crops.

With respect to total acres improved since 1972, no relationship could be found. This concurs with the analysis presented earlier in the paper. Therefore, while we cannot argue that the subsidized nature of Saskatchewan crop insurance has not encouraged the cultivation of marginal lands, the results presented in this paper do not support the conclusion that crop insurance by itself was an important factor. One could argue that factors like changes in relative prices and other government programs have dominated the effects of crop insurance. This is supported by the research of Van Kooten and Schmitz (1989).

Table 8. Empirical results of the probit model[a]

	Region			
Variable	Chaplin	Moose Jaw	Meadow Lake	Melfort
Constant		0.69 (1.36)	0.6 (1.27)	1.66 (1.23)
ST/AC	-0.71 (-0.91)			-1.48 (-1.10)
ACBR	--	--	--	--
VAIP	0.038 (1.82)	-0.028 (-1.09)	-0.016 (1.44)	0.021 (1.10)
SNTP		0.82 (1.75)	-0.78 (-1.13)	
CHFL		1.18 (2.55)		
STCP	-0.4 (-1.16)			
LEGM			-0.79 (-1.59)	-0.47 (-1.32)
CPNO				-0.21 (-1.52)
LDUM	0.54 (1.93)	-1.28 (-2.90)	-0.44 (-1.20)	
NOOB	64	64	60	58
Log Likelihood	-38.77	-28.71	-35.76	-36.92

[a]T statistics are in parentheses. The dependent variable in the probit is 1 for a participator in crop insurance and 0 otherwise.

research of Van Kooten and Schmitz (1989).

A comparison of the level of variable inputs (VAIP) used by participators and non-participators does not yield conclusive results either. In some regions, the probit shows that participators use more inputs than non-participators, while in others the opposite result appears. Therefore, if one can argue that participators are more risk averse than non-participators (Weisensel and Schoney 1990), the results seem to support the hypothesis that crop insurance tends to reduce the differences in input use between risk averse and risk neutral producers.

When one examines other land use variables, it appears that participators are more likely to stubble snow trap and chemical fallow but are less likely to use strip cropping techniques.[24] In Meadow Lake and Melfort, those farmers who seed more crops and include a legume in their rotation tend not to participate. This is not a surprising result since these variables are measuring other sources of risk reducing farm practices.

Finally, livestock, an enterprise which uses marginal land, has mixed results as well. In Moose Jaw and Meadow Lake, participators are less likely to have livestock income, confirming the hypothesis that participation in crop insurance favors land used in grain production. However, in Chaplin, the results suggest that participators are more likely to have livestock income. One possible explanation for the contrast in the reported parameters is that in the Chaplin area a substantial amount of marginal pasture land exists. Therefore, cow calf enterprises are prevalent on many farms in this region.

[24]Chemical fallow is a substitute for regular tillage fallow. The practice employs herbicides to control weed growth during the fallow period. Stubble snow trapping is a practice used to trap more snowfall on stubble fields over the winter months. It is accomplished by cutting stubble at alternating heights.

13.8 Conclusions

In this paper, we examine the impact of crop insurance on land use decision in the province of Saskatchewan using a number of approaches. First, we present a conceptual model which suggests how some basic farm management decisions may change once an individual participates in crop insurance. Second, we provide empirical evidence that suggests that land use decisions do change marginally as a result of crop insurance. Using a Likert scale, we find that a number of participators in crop insurance use more inputs and stubble crop more frequently than if the insurance program did not exist. In addition, using the same approach, we find that crop insurance has an influence upon the number of crops some participators grow as well as the fields they allocate to specific crops. The number of farmers influenced by the program is greater in the two southern areas (Chaplin and Moose Jaw) than the two northern areas. This is not surprising since the expected value of insurance tends to be greater in the southern cropping areas of the province (Weisensel, Furtan and Schmitz 1990).

A second component of the paper examines differences between the land use variables of participators and non-participators. This analysis provides a comparison of the two groups of producers after the participators have already adjusted to the insurance program. Therefore, if the participators in crop insurance tend to be more risk averse than non-participators, participation in insurance should make the farming practices of the participator appear more like the non-participator (for instance, input use and stubble cropping). The probit model presented appears to support this conjecture.

With respect to the cultivation of marginal land, it appears that the crop insurance program has had little measurable impact. However, since crop insurance is a subsidy to grain production, it is difficult to conclude that crop insurance has not encouraged the cultivation of marginal land. One could argue that changes in relative prices between grain and livestock as well as other government programs over the last 20 years have had more important impacts on marginal land use than has crop insurance. This concurs with recent research completed by Van Kooten and Schmitz (1989).

Acknowledgements. Subject to the usual qualifier, the authors would like to thank Pauline Molder and Julia Taylor for their many helpful comments and suggestions. Funding for this research was provided by the Agriculture Development Fund, Saskatchewan Department of Agriculture and Food.

References

Agriculture Canada (1989) Federal-provincial crop insurance review: final discussion paper

Anderson JR, Dillon JL, Hardaker B (1977) Agricultural decision analysis. Iowa State Univ Press, Ames, Iowa, 344 pp.

Just RE, Pope RD (1978) Stochastic specification of production functions and economic implications. J Econometrics 7:67-86

Nelson C, Loehman E (1987) Further toward a theory of agricultural insurance. Amer J Agr Econ 69:523-531

Newberry DMG, Stiglitz JE (1981) The theory of commodity stabilization—a study in the economics of risk. Oxford Univ Press, Oxford

Rennie DA (1986) Soil degradation: a western perspective. Canad J Agr Econ: Proceedings Issue 33:19-29

Rosaasen K, Eley R, Lokken, J (1990) Federal government relief programs for grain farmers: rewards for the late adjusters? Paper presented at the Soils and Crops Workshops. Dept of Agric Econ, Univ Saskatchewan, pp. 354-379

Saskatchewan Crop Insurance Corporation (selected years) Saskatchewan crop insurance annual report. Government of Saskatchewan

Saskatchewan Agriculture (1988) Agricultural statistics 1987, Economics Statistics Section

Schoney RA, Thorson T, Weisensel WP (1988) 1988 results of the Saskatchewan top management workshops. Dept Agricultural Economics, Univ Saskatchewan, 78 pp.

Van Kooten GC, Schmitz A (1989) Socioeconomic evaluation of incentives to promote waterfowl habitat: prairie pothole project extensive survey results. Report prepared for Saskatchewan Dept of Parks, Recreation and Culture

Walizer MH, Wienir PL (1978) Research methods and analysis: searching for relationships. Harper and Row, Publishers, New York

Weisensel WP, Furtan WH, Schmitz A (1990) The Saskatchewan all-risk crop insurance program: an evaluation of participation, land use and technology adoption. Technical bulletin, Dept Agricultural Economics, Univ of Saskatchewan

Weisensel WP, Schoney RA (1989) An analysis of the yield-price risk associated with specialty crops. West J Agric Econ 14(2):293-299

Weisensel WP, Schoney RA (1990) An analysis of producer participation in government insurance programs: are risk preferences an important factor in explaining participation? Technical paper, Dept of Agricultural Economics, Univ Saskatchewan

Chapter 14

Providing Catastrophic Yield Protection Through a Targeted Revenue Program

J.W. GLAUBER[1] and M.J. MIRANDA[2]

Much of the debate surrounding the 1990 Farm Bill has focused on the best means of providing producers with income protection in the event of a widespread yield shortfall. The debate is not new. Prior to 1980, the U.S. Department of Agriculture provided disaster assistance mainly through direct cash payments. Because of strong criticism that the disaster payment program was too expensive, restrictive in scope (only program crops were eligible for payments), and encouraged production in high-risk areas, the Federal Crop Insurance Act of 1980 was enacted to replace the disaster payment program, with crop insurance as the primary means of providing disaster protection to farmers. To encourage participation in the crop insurance program, producers were offered direct premium subsidies of up to 30 percent, and additional indirect subsidies.

Despite an expansion of the federal crop insurance program in recent years to include additional crops and counties, participation in the program has remained low. In 1988, less than 30 percent of eligible acreage was enrolled. While estimated participation in 1989 is 44 percent, much of the increase was due to the 1988 Disaster Assistance Act which required producers to purchase crop insurance to qualify for 1988 disaster payments (Glauber, Harwood and Miranda 1989).

The crop insurance program has been costly. The U.S. General Accounting Office (1989) estimates that over the period

[1]Consumer Economics Division, Economic Research Service, U.S. Department of Agriculture, Washington, DC, USA.

[2]Department of Agricultural and Resource Economics, The Ohio State University, Columbus, Ohio, USA.

1981-88, costs for the program averaged nearly $500 million annually. Excess losses (indemnities minus total premiums, including producer subsidies) accounted for almost 47 percent of total costs. Moreover, low participation in the crop insurance program has contributed to the passage of supplemental ad hoc disaster assistance legislation in five of the past nine years. Total costs for ad hoc disaster assistance and crop insurance combined averaged over $1.1 billion annually from 1981-88, exceeding $6 billion in 1988 and 1989 alone.

Ad hoc disaster assistance legislation has significantly undermined the 1980 Federal Crop Insurance Act's goal of establishing crop insurance as the primary source of disaster risk management for farmers. Many producers forego purchasing crop insurance in the belief that the federal government will provide supplemental ad hoc disaster relief. The problem is self-perpetuating. Low participation makes it all the more likely Congress will enact disaster legislation when widespread disasters occur.

The failures of the current crop insurance program raise serious concerns about whether a fiscally responsible crop insurance program could attract sufficient participation to reduce the demand for supplemental ad hoc disaster legislation. Citing these concerns, the Bush Administration recommended eliminating federal crop insurance and replacing it with a standing disaster assistance program which, unlike its predecessor in the 1970s, would indemnify producers for crop losses only when catastrophic crop yield losses are experienced at the county level. Critics have charged that the program under the coverage levels proposed by the Administration would provide producers with insufficient protection (Hourigan, Skees and Barnett 1990).

In order to provide overall revenue risk protection, the federal government has also operated a deficiency payments program, designed to provide price risk protection, concurrently with crop insurance and ad hoc disaster assistance programs. Under the current deficiency payment program, producers receive payments proportional to the difference between a target price and the higher of the market price or the nonrecourse loan rate. In the event of a widespread crop disaster, such as occurred in the Corn Belt in 1988 or the Southern Plains in 1989, market prices rise, thus reducing the size of the deficiency payments, and leaving farmers with little compensation for yield losses.

As an alternative to the current combination of target price deficiency payments, crop insurance, and ad hoc disaster assistance, we propose a modified deficiency payments program in which the deficiency payment would be based on the difference between a target revenue and the average revenue in the region. Such a program would leave intact individual incentives to obtain the maximum yield and the highest price for a crop, thus minimizing potential moral hazard problems. Because of the negative correlation between aggregate yield and price, revenue deficiency payments would be higher than under the current program when yields are low, and lower when yields are high. The net effect is to stabilize both total government deficiency payments and net income for individual producers, thereby reducing the need for ad hoc disaster assistance.

In the following section, we present a regional model of the U.S. corn market. Using numerical simulation methods, we calculate the distribution of producer revenues and federal budget outlays under the current target price deficiency payment program. We then derive the distributions for deficiency payment programs that base payments on the difference between a target revenue and either the average regional revenue or the average national revenue. The analysis suggests that significant revenue protection can be gained at no additional cost to the government by using a target revenue program rather than a target price program.

14.1 The Model

Under the current federal income support program, participating producers receive a per-acre deficiency payment that is calculated by multiplying the national deficiency payment rate by the individual program yield. The deficiency payment rate is the difference, if positive, between the target price P^T and the greater of the market price P and the loan rate P^L. Denoting the average program yield in region i by Y^*_i, the average per-acre deficiency payment received by participating producers in region i under the current program, if the market price were P, would be:

$$Def^1 = Max(0, P^T - Max(P, P^L)) \cdot Y^*_i. \quad (1)$$

We examine two alternative methods for calculating per-acre deficiency payments based on a target revenue rather than on a target price. Under a national target revenue program, participating producers would receive a basic per-acre deficiency payment equal to the difference, if positive, between the national target revenue R^T and participant national average revenue, which equals the maximum of the market price and the loan rate times the national average yield Y. In order to account for regional differences in expected yields, the basic payment is adjusted by a factor equal to the ratio of the average regional program yield Y^*_i and the national program yield Y^*. In other words, the average per-acre deficiency payment received by participating producers in region i under a national target revenue program would be:

$$Def^2 = Max(0, R^T_i - Max(P, P^L) \cdot Y) \cdot Y^*_i/Y^*. \quad (2)$$

Under a regional target revenue program, participating producers in region i would receive a per-acre deficiency payment equal to the difference, if positive, between the regional target revenue R^T_i and the participant regional average revenue, which equals the maximum of the market price and the loan rate times the regional average yield Y_i:

$$Def^3_i = Max(0, R^T_i - Max(P, P^L) \cdot Y_i). \quad (3)$$

In order to estimate the distributions of total government deficiency payments and regional per-acre revenues under the alternative deficiency payment schemes, we apply stochastic simulation techniques. In our analysis, we employ a fully parameterized annual model of the U.S. corn market divided into six regions: Illinois, Indiana, Iowa, Kentucky, Ohio, and the rest of the United States. Table 1 gives the 1989 acreage levels and program yields used in our simulations and the target revenues selected for the regional and national target revenue programs. In our simulations, we set the national target price at \$2.84 per bushel and the national loan rate at \$1.65, their 1989 values. The total demand function is formulated as the sum of domestic and export demand functions, both of which are specified in Cobb-Douglas form expressing quantity demanded in billions of bushels, and price in 1989 dollars per bushels. The constant term and elasticity for the domestic demand function are 6.615 and -0.3, respectively; the constant term and elasticity for the export demand function are 3.392 and -0.9, respectively. Differences among regional market prices due to basis are ignored.

Since the degree of systematic variation in regional yields is an important determinant of both the expected level and variability of government deficiency payment expenditures, care is taken to replicate in our model the observed covariances between

Table 1. Planted acreage, program yield, and target revenue, by region

Region	Planted Acreage	Program Yield	Target Revenue
	Million Acres	Bushels per Acre	Dollars per Acre
Illinois	10.8	117	356.6
Indiana	5.2	110	343.9
Iowa	12.3	117	350.8
Kentucky	1.2	95	289.4
Ohio	2.9	110	337.9
Other	32.4	94	301.0
U.S.	64.8	104	324.4

the regional yields and the national yield. We accomplish this by orthogonally projecting the logarithm of the regional yield onto the logarithm of the national yield, arriving at the identity:

$$y_i = y_i + \beta_i \cdot (y - \bar{y}) + \in_i, \qquad (4)$$

where y_i is the logarithm of the yield in region i, y is the logarithm of the national yield, and $\bar{y}_i$ and $\bar{y}$ are their respective expectations; further,

$$\beta_i = \mathrm{Cov}(y_i, y)/\mathrm{Var}(y) \qquad (5)$$

is a factor that measures the sensitivity of the regional yield to systemic variations in national yield and $\in_i$ is a residual stochastic term which, by construction, is uncorrelated with the log of the national yield. The identity allows a decomposition of yield variability in region i into systemic and residual components as given by:

$$\mathrm{Var}(y_i) = \beta_i^2 \cdot \mathrm{Var}(y) + \mathrm{Var}(\in_i). \qquad (6)$$

Table 2 gives the expected yield, yield variability, yield beta, and percent systemic and nonsystemic yield variation by region.

14.2 Simulation Results

Simulated expected per-acre revenues and total deficiency payments for the 1989 corn marketing year are given in Table 3. Since we ignore differences between regional prices and the national price, regional differences in simulated market revenues in the absence of deficiency payments are due exclusively to regional variations in yield distributions. Expected per-acre revenues under the three alternative deficiency payments programs, on the other hand, further reflect the underlying regional and national program yields and the level of the national target price, the national target

Table 2. Expected yield, yield variability, and decomposition of yield variability, by region

Region	Expected Yield	Standard Deviation of Yield	Beta	Yield Variation	
				Systemic	Residual
	Bushels per Acre			Percent	
Illinois	132.6	22.4	1.52	89.5	10.5
Indiana	129.0	18.7	1.22	85.4	14.6
Iowa	128.9	17.5	1.20	86.3	13.7
Kentucky	107.4	19.5	1.28	53.7	46.3
Ohio	125.2	14.6	0.90	63.9	36.1
Other	113.5	9.3	0.74	88.1	11.9

revenue, or the regional target revenue. By design, the revenue targets (see Table 1) were chosen so as to assure that the expected revenues in each region, and thus the expected government expenditures, were the same under all three deficiency payment

Table 3. Expected per-acre revenue, by region, and total deficiency payments under alternative programs

Region	No Deficiency Payments	Target Price	National Target Revenue	State Target Revenue
	Dollars per Acre			
Illinois	240.1	356.7	356.7	356.7
Indiana	234.3	343.9	343.9	343.9
Iowa	234.2	350.8	350.8	350.8
Kentucky	194.9	289.5	289.5	289.5
Ohio	228.3	337.9	337.9	337.9
Others	207.4	301.0	301.0	301.0
	Billions of Dollars			
Total Deficiency Payments	0.00	6.72	6.72	6.72

Table 4. Variability of per-acre revenue by region and variability of total deficiency payments under alternative programs

Region	Without Deficiency Payments	Target Price	National Target Revenue	State Target Revenue
	Percent Coefficiency of Variation			
Illinois	10.02	11.19	5.24	0.06
Indiana	9.27	9.29	4.17	0.01
Iowa	8.91	8.92	3.58	0.00
Kentucky	14.27	11.94	8.39	0.31
Ohio	10.63	7.39	4.72	0.15
Others	9.63	4.58	3.19	0.08
	Billions of Dollars			
Total Deficiency Payments	0.00	28.61	16.41	16.37

programs. This allows direct comparison among the alternative deficiency payments programs solely on the basis of their ability to stabilize regional per-acre revenues and total government costs.

Table 4 shows the variability of per-acre revenue and government expenditures for the three deficiency programs. The variability of revenues without deficiency payments is shown for comparison purposes. Note that for some regions, the target price deficiency payment program destabilizes per-acre revenues when compared to a market without deficiency payments. At the aggregate level, price and yield are negatively correlated. If aggregate yield falls, prices rise, and vice versa. Price responsiveness thus provides a "natural hedge" against the revenue shortfalls that might otherwise result from yield shortfalls. The target price deficiency payment program stabilizes the effective price received by the producers near the target price, thereby removing the natural hedge that exists between price and yield revenue (Grant 1985, Miranda and Helmberger 1988). Since the stabilizing effect of the natural hedge on regional revenues is strongest for regions possessing a high degree of systemic yield variation, it is those regions that suffer revenue destabilization under a target price program.

Both the national and regional target revenue programs base deficiency payments on revenue shortfalls, rather than price shortfalls. Because the target revenue programs do not destroy the natural hedge between price and yield, but rather exploit it, revenues under these programs are more stable than under the target price program. Because of exogenous factors affecting acreage supply and export and domestic demands in 1989, and the relatively high levels at which the regional and national target revenues are set, the probability of a regional or national revenue exceeding its target is negligible. Under these circumstances, the national target revenue program removes all but the residual yield risk component of regional revenue variability (see Table 2), thereby substantially stabilizing the regional revenue. The regional target revenue program further removes the residual yield risk, thereby stabilizing regional per-acre revenue perfectly.

Fig. 1 shows the cumulative probability function of per-acre revenue for Illinois under a target price program, a national target revenue program, and a regional target revenue program. While the mean revenue is equal for all three programs ($365 per acre), revenue is much less dispersed under the regional target revenue program than under a national target revenue program, which, in turn, is much less dispersed than under a target price program. Similar figures obtain for the other regions.

Table 5 shows the amount of downside revenue risk protection afforded by the three programs. The probability of revenue falling below 90 percent of the expected revenue under the target price program is greater than under either the regional or national target revenue program. The probability of a comparable revenue shortfall under a national target revenue program is small but not zero, reflecting the possibility, albeit rare, of a localized disaster that adversely affects yields regionally at a time when the rest of the country enjoys average or above average yields. Of

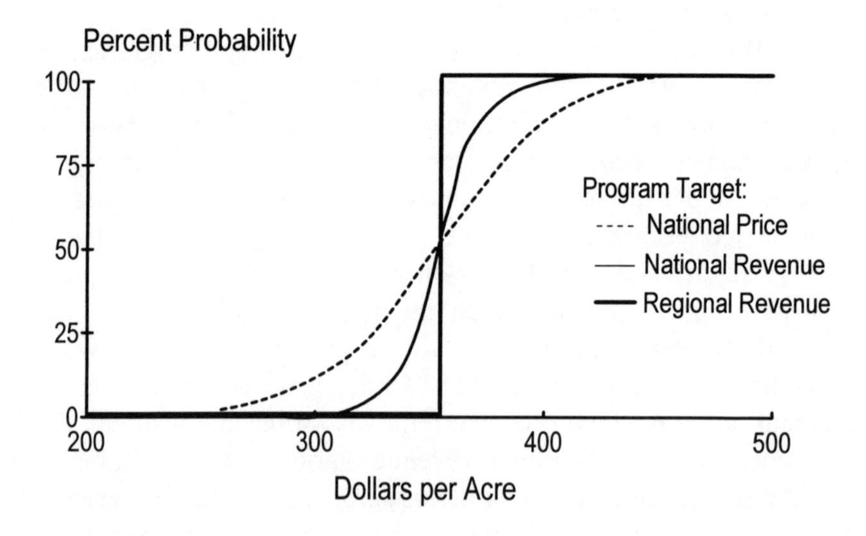

Fig. 1. Cumulative distribution regional per-acre revenue: Illinois

Table 5. Probability that regional per-acre revenue will fall below 90% of regional target revenue under alternative programs, by region

Region	No Deficiency Payments	Target Price	National Target Revenue	State Target Revenue
	Percent Probability			
Illinois	99.2	18.8	4.0	0.0
Indiana	99.7	14.8	1.3	0.0
Iowa	99.9	14.1	0.7	0.0
Kentucky	98.3	21.3	13.3	0.0
Ohio	99.3	10.0	2.1	0.0
Others	98.7	1.9	0.2	0.0

course, the probability of the regional yield falling below 90 percent of average under the regional target revenue is zero, reflecting the absolute floor put on regional revenue by such a program. Table 5 suggests that both the regional and national target revenue programs would substantially reduce downside revenue risk and thus the demand for supplemental ad hoc disaster assistance.

Because revenues are less variable under the target revenue programs than under a target price program, deficiency payments under the target revenue programs are more stable as well (Table 4). Fig. 2 shows the cumulative distribution function of deficiency payments under all three alternative deficiency payments programs. In our simulations, the distribution of total deficiency payments under the two target revenue programs differed by such small amounts as to make their distributions indistinguishable graphically. Under a target price program, per-bushel payments may range from near zero to full payment when the market price is below the loan rate. For the 1989 corn crop, the probability of full payment under the target price program was almost 50 percent. This is evident from a vertical singularity in the cumulative distribution of payments under the target price program beginning

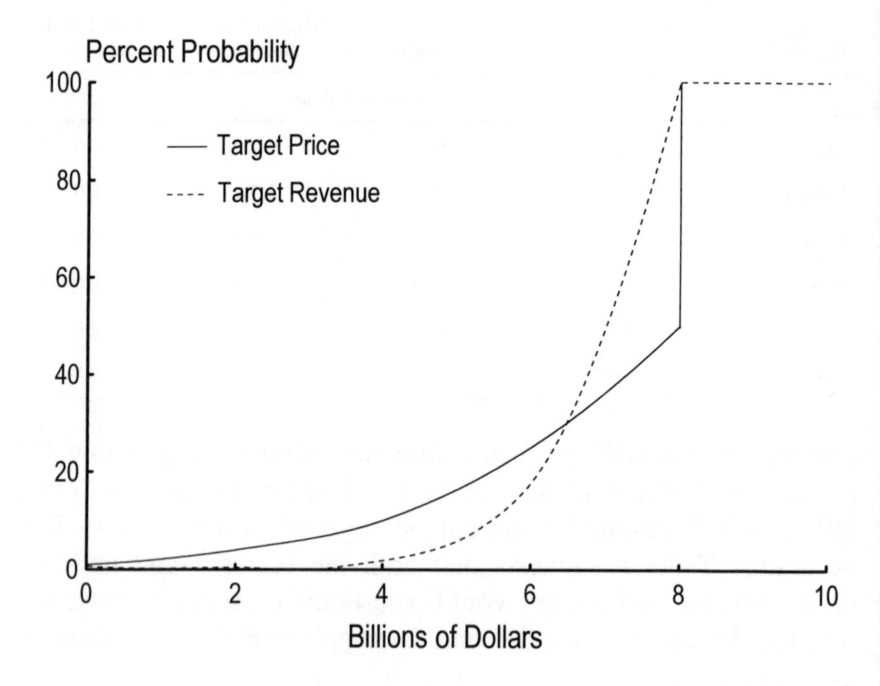

Fig. 2. Cumulative distribution of total deficiency payments

at $8 billion, the aggregate full payment level.

Payments under the national and regional target revenue programs are much less variable and smoothly distributed. The correlation between price and yield reduces the relative variability in deficiency payments. In general, for years when aggregate U.S. yields are below program yields, the target revenue programs tend to pay producers more than under the target price program. During years when yields exceed program yields, payments are significantly less than under the target price program. This reduction in outlay variability would lessen the probability of sequestration under the Gramm-Rudman-Hollins Deficit Reduction Act.

14.3 Conclusions

In this paper we have analyzed two program alternatives to the current target price deficiency payments program. The alternative programs base deficiency payments on shortfalls in per-acre revenues rather than price. We have shown that under the target revenue programs, producers would receive the same revenue on average as under the current program for the same cost to the government, but with significant reduction in the variabilities of government expenditures and regional producer revenues. In particular, under a target revenue program, producers would receive substantially greater downside revenue risk protection, potentially eliminating the need for supplemental ad hoc disaster legislation.

The deficiency payments programs considered in this paper are pure income transfer rather than insurance programs. As an alternative, producers could be charged a premium to pay for some portion of the protection they receive. Since revenues would be calculated at the regional level rather than at a farm level, adverse selection and moral hazard problems would be minimized. Demand for such insurance would be tied to the level of protection offered to the individual producer, which is, in turn, determined by the correlation between the producer's yield and the regional yield.

Extensions of the model employed in this paper include development of acreage supply equations that incorporate risk response. The reduction in revenue variability under a target revenue program would likely encourage increased production by risk-averse participants and entry into the program by nonparticipating producers, thus increasing aggregate government outlays; these effects would have to be captured. Another extension would be to analyze the effects of target revenue programs on individual producer revenue risk, rather than on the average regional revenue variability; this would require empirical estimates of the relation between farm yields and regional yields.

References

Glauber JW, Harwood JL, Miranda MJ (1989) Federal crop insurance and the 1990 farm bill: an assessment of program options. Economic Research Service, U.S. Department of Agriculture. Staff report no. AGES89-45

Grant D (1985) Joint output and price uncertainty. Amer J Agric Econ 67:630-635

Hourigan JD, Skees JR, Barnett BJ (1990) An evaluation of the administration's disaster assistance program. Unpublished manuscript, Univ Kentucky

Miranda MJ, Helmberger PG (1988) The effects of price band buffer stock programs. Amer Econ Rev 78:46-58

U.S. General Accounting Office. Disaster assistance: crop insurance can provide assistance more effectively than other programs. Report to the Chairman, Committee on Agriculture, U.S. House of Representatives, Washington, D.C., GAO/RCED-89-211

Author Index

Subject Index